BY ROBERT KURSON

Shadow Divers
Crashing Through
Pirate Hunters
Rocket Men

Praise for *Rocket Men*

"Kurson's first-rate account of this remarkable spaceflight starts by reminding us what a gamble it was. There are many pieces to the Apollo 8 story, but Kurson brings them together effortlessly." —*USA Today*

"This nonfiction page-turner reads like a novel, complete with plot twists and on-the-edge-of-your-seat anxiety, with historical context skillfully interwoven into the personal stories of three men who orbited the Moon. [A] compelling, meticulously-researched book by a master storyteller."
—*Forbes*

"*Rocket Men* is one of the best descriptions of an Apollo flight I've ever read. Kurson's account let me relive all the thrilling moments from this historic mission. I highly recommend this book to the 'space cadet'— which I consider myself to be. *Rocket Men* is as good as it gets."
—CHRIS KRAFT, director of Apollo flight operations, NASA

"Engrossing . . . Kurson's conception-to-splashdown reporting had the cooperation from the astronauts and their wives, giving him invaluable details of what happened inside the astronauts' capsule and in their homes below." —Associated Press

"Apollo 11 grabbed the glory, but Apollo 8 was the mission that proved humans could travel to the Moon. . . . This is the story of their mission, told in cinematic detail." —NBC News

"Kurson's evocative writing places Apollo 8 into historical perspective and allows us to vicariously experience the launch of the Saturn V rocket and the awe felt by the first men to leave [Earth]." —*Science*

"This is a great story . . . The best book I've read this year."
—JIM BRIDENSTINE, head of NASA

"Spellbinding." —*American History* magazine

"Gripping . . . The story of the dangerous mission that laid the ground for the Moon landing has not been told in such detail until now."

—*The Economist*

"Breathing life into history is a specialty of Robert Kurson's. . . . [He] never allows his exhaustive research to get in the way of the story, crafting remarkable nonfiction adventures that put readers at the center of the action. *[Rocket Men]* does that and more."   —*Mountain Times*

"Absolutely riveting . . . A gripping tale well told."   —*Booklist*

"Kurson tells the tale of Apollo 8 with novelistic detail and immediacy, expertly capturing the urgency and suspense behind the mission that gave America the lead in the Space Race."   —ANDY WEIR

"*Rocket Men* is a timely and thrilling reminder of a heroic American achievement. It has it all—suspense, drama, risk, and loving families. We could use those days again."   —TOM BROKAW

"This is the story of the most consequential and daring voyage since those in the era of Columbus. The tale is told with the care and clarity, and the heart-banging drama, that Robert Kurson's legion of readers have come to expect from him. A flat-out terrific book."   —SCOTT TUROW

"As a Rocket Woman myself, I know very well the story of Apollo 8, and yet I couldn't put this book down. Kurson presents not only the challenges, risks, ambition, and success of Apollo 8, but a story of human spirit."   —NICOLE STOTT, NASA ISS and Shuttle Astronaut

"Apollo 8 was perhaps the most aggressive, audacious, and dangerous mission devised by NASA in its race with the Soviet Union to land humans on the Moon. *Rocket Men* tells the thrilling story of this historic mission through the eyes of its remarkable crew, three men who had the admiration and support of the entire astronaut corps."

—MAJ. GEN. JOE H. ENGLE (Ret.), X-15 test pilot and NASA astronaut

ROCKET
MEN

The Daring Odyssey of Apollo 8
and the Astronauts Who Made
Man's First Journey to the Moon

ROBERT KURSON

RANDOM HOUSE
NEW YORK

2019 Random House Trade Paperback Edition

Copyright © 2018 by Robert Kurson

All rights reserved.

Published in the United States by Random House,
an imprint and division of Penguin Random House LLC, New York.

RANDOM HOUSE and the HOUSE colophon are registered
trademarks of Penguin Random House LLC.

Originally published in hardcover in the United States by Random House,
an imprint and division of Penguin Random House LLC, in 2018.

LIBRARY OF CONGRESS CATALOGING-IN-PUBLICATION DATA
Names: Kurson, Robert, author.
Title: Rocket men: the daring odyssey of
Apollo 8 and the astronauts who made man's
first journey to the Moon / Robert Kurson.
Description: First edition. | New York: Random House, [2018] |
Includes bibliographical references and index.
Identifiers: LCCN 2017009386 | ISBN 9780812988703 (hardback)
| ISBN 9780812988710 (trade paperback) | ISBN 9780812988727 (ebook)
Subjects: LCSH: Project Apollo (U.S.) | Space flight to the Moon.
| Apollo 8 (Spacecraft)
Classification: LCC TL789.8.U6 A5438 2018 | DDC 629.45/4—dc23
LC record available at lccn.loc.gov/2017009386

Printed in the United States of America on acid-free paper

randomhousebooks.com

2 4 6 8 9 7 5 3 1

Book design by Elizabeth A. D. Eno
Frontispiece photographs courtesy of NASA
Images page 342–343 adapted from NASA/Mitch Lopata

For Amy, my best friend in all the universe

For Nate and Will, star sailors

CONTENTS

ROCKET MEN

Prologue: Countdown

●

THREE ASTRONAUTS ARE STRAPPED INTO A SMALL SPACECRAFT thirty-six stories in the air, awaiting the final moments of countdown. They sit atop the most powerful machine ever built.

The Saturn V rocket is a jewel of the National Aeronautics and Space Administration, a vehicle that will generate the energy of a small atomic bomb. But it has never flown with men aboard, and it has had just two tests, the most recent of which failed catastrophically just eight months earlier. The three astronauts are going not merely into Earth orbit, or even beyond the world altitude record of 853 miles. They intend to go a quarter of a million miles away, to a place no man has ever gone. They intend to go to the Moon.

Beneath them, the United States is fracturing. The year 1968 has seen killing, war, protest, and political unrest unlike any in the country's history, from the assassinations of Martin Luther King, Jr., and Robert Kennedy to the unraveling of Vietnam to the riots in Chicago. Already, *Time* magazine has named THE DISSENTER its Man of the Year.

As the countdown begins, there are engineers and scientists at NASA who question whether the crew will ever return. Even the astronauts are realistic about their chances of surviving the flight, an operation riskier than anything the American space agency has ever attempted. One of them has recorded a final goodbye to his wife, to be played in the event he doesn't return.

In August, this mission did not exist. Nearly everything that has gone into its planning—the training, analysis, calculations, even the politics—has been rushed to the launchpad in a fraction of the time ordinarily required. If anything goes wrong, public opinion—and the will of the United States government—might turn against NASA. The fate of the entire space program hangs on the crew's safe return.

As the moment of launch draws near, one of the astronauts spots a mud dauber wasp building a nest on the outside of one of the spacecraft's tiny windows. Back and forth the insect moves, grabbing mud and adding to its new home. The astronaut thinks, "You are in for a surprise."

Vapors begin to spew from around the base of the giant rocket. Less than a minute remains before lift-off. When the five first-stage engines ignite, they will deliver a combined 160 million horsepower. In the final few seconds, a typhoon of flames unfurls to either side. Beneath the astronauts, it is not just the launchpad that begins to shake, but the entire world.

Chapter One

●

DO YOU WANT TO GO TO THE MOON?

August 3, 1968—Four months earlier

AS HE SAT ON A BEACH IN THE CARIBBEAN, A QUIET ENGINEER named George Low ran his fingers through the sand and wondered whether he should risk everything to win the Space Race and help save the world.

At forty-one, Low was already a top manager and one of the most important people at NASA, in charge of making sure the Apollo spacecraft was flightworthy.

Apollo had a single goal, perhaps the greatest and most audacious ever conceived: to land a man on the Moon and return him safely to Earth. In 1961, President John F. Kennedy had committed the United States to achieving this goal by the end of the decade. Never had a more inspiring promise been made to the American people—or one that could be so easily verified.

Now, Kennedy's end-of-decade deadline was in jeopardy. Design and engineering problems with the lunar module—the spidery landing craft that would move astronauts from their orbiting ship to the lunar surface and back again—threatened to stall the Apollo program and put Kennedy's deadline, just sixteen months away, out of reach. And that led to another problem. Every day that Apollo languished, the Soviet Union moved closer to landing its own crew on the Moon. And that mattered. The nation that landed the first men on the Moon would score the ultimate victory in the years-long Space Race between the two superpowers, one from which the second-place finisher might never recover.

For months, NASA's best minds had worked around the clock to fix the issues with the lunar module, but the temperamental and complex landing craft only fell further and further behind schedule. By summer, many at the space agency had abandoned hope of making a manned lunar landing by the end of the decade.

And then Low had an idea.

It had come to him just a few weeks before he'd arrived at this beach, and it was wild, an epiphany, a dream. It was also dangerous, risky beyond anything NASA had ever attempted. But the more Low thought about it, the more he believed it could keep the Apollo program moving and save Kennedy's deadline—and maybe even beat the Soviets to the Moon.

Low inhaled the fresh, salty air and tried to push space travel out of his thoughts. At home, his mind burned nonstop with ideas, formulae, trajectories. Now he needed a break, and it should have been easy to find one in this tropical paradise. About the only reminder of America was the local newspaper, which told of the Newport Pop Festival in Costa Mesa, California, where more than a hundred thousand music fans were expected, and brought word of potential protests at the coming Democratic National Convention in Chicago. It had been an explosive year already, with assassinations, riots, and violence. A quiet beach was just where a man like Low needed to be.

But Low could not relax. He walked the beach, looking out

over the ocean toward Moscow and the Moon, thinking, imagining, America and the world on fire behind him.

Five days after Low returned from vacation, a serious man with an oversized head went to work inside a giant assembly plant in Downey, California. His mission: to build a machine from the future that would help make the world safe for democracy.

Over and over, astronaut Frank Borman opened and closed the hatch on the Apollo command module, a cone-shaped capsule made to fly a three-man crew to the Moon. He'd already certified that the hatch worked, then certified it again, but he would not stop pushing on it, making sure it opened, no matter what.

Nearby, Borman's two crewmates, Jim Lovell and rookie Bill Anders, got ready to test the hundreds of dials, switches, levers, lights, and gauges that made the command module work. The spacecraft was small, measuring just eleven feet tall and thirteen feet wide at its base, but every inch of it had been designed by Borman and others to be impervious to a galaxy of deadly forces.

A nearby transistor radio played Top 40 music, which caught Borman's ear.

"That's a pretty slick song," Borman said. "Who's the fella singing it?"

"That's the Beatles, Frank," Lovell said, laughing.

Borman preferred the standards. As a kid, he'd memorized the lyrics to all the great Western songs played on the radio in Arizona. He could still sing "Cowboy Jack"—a ditty that dated to the nineteenth century—but didn't dare start, because he knew Lovell and Anders would insist that he sing it to the end.

Borman stuck to classic films, too. Alone among astronauts, it seemed, he hadn't bothered to see *2001: A Space Odyssey,* the new Stanley Kubrick film released in April that showed men flying to the Moon. That stuff was science fiction, Borman told his colleagues; America had real people to get to the Moon.

Borman and his crewmates knew that the lunar module was troubled and behind schedule. But until designers and engineers could make the fixes, these astronauts could do little more than make certain that the command module was perfect. So they climbed inside their spacecraft and began testing it, pushing the command module mercilessly, because that's what outer space would do to it, too.

And then the phone rang.

Smart people knew better than to bother Borman at work. But the man on the line went back a long way with Borman. And he said it was urgent.

Donald Kent "Deke" Slayton was in charge of managing astronaut training and choosing crews for manned space missions. If an astronaut flew on board a NASA spacecraft, it was because Slayton had chosen him to go.

When Borman heard who was calling, he wriggled out of the capsule and grabbed an extension.

"Deke, I'm in the middle of a big test here," he said.

"Frank, I need you back in Houston."

"Talk to me now."

"No, I can't talk over the phone. It's gotta be in person. Grab an airplane and get to Houston. On the double."

Borman grimaced—America did not have time for nonsense and delays—but Slayton was in charge, and NASA, no matter its official designation as a civilian organization, was a military operation to Borman, so he took his orders. Poking his head back inside the spacecraft, he told his partners, "You guys are stuck with the module. I've gotta go back to Houston."

Borman grabbed his rental car, drove to Los Angeles International Airport, and hopped into a T-38 Talon, a two-seat twin-engine supersonic jet used by astronauts for training, commuting, and even some fun, and pointed it toward Texas. At forty, he still looked every bit the West Point cadet: sandy blond near-crewcut, square jaw and chin set for combat, arched eyebrows that seemed a radar for anything askew. Even his head was military issue, all right angles and slightly larger than life, a feature that had earned him the childhood nickname Squarehead.

Borman couldn't imagine why he was needed in Houston, and so suddenly. He was commander of Apollo 9, the third of four manned test flights NASA planned before it would attempt to land on the Moon. Apollo 9 was to be a basic mission—orbit Earth, test the spacecraft, come home. It wasn't scheduled to launch for another six months. Still, Borman knew he hadn't been summoned for nothing. The last time he'd received a "drop everything" call had been the darkest day in NASA's history.

It had happened about a year and half earlier, on January 27, 1967, when a fire broke out in the spacecraft during a simulated countdown on the launchpad in Florida. The Apollo 1 rehearsal should have been safe and routine for the three astronauts inside, who were preparing for the actual flight about four weeks later. But a spark occurred in the electrical system and the men were trapped as the sudden fire spread in pure oxygen. Even Ed White, the strongest of all NASA's astronauts, couldn't muscle open the command module's hatch as flames spread through the spacecraft.

Borman had been enjoying a rare break with his family at a lakeside cottage near Houston, where they lived, when Slayton's call came in that day.

"Frank, we've had a bad fire on Pad Thirty-four and we've got three astronauts dead—Gus Grissom, Ed White, and one of the new boys, Roger Chaffee. Get to the Cape as quick as you can; you've been appointed to the investigative committee."

The news stunned Borman, who considered Ed White the brother he'd never had. And it devastated Borman's wife, Susan, who counted Pat White among her best friends. Borman told Slayton he'd fly to Florida right away but first needed to stop at the Whites' home in Houston.

When he and Susan arrived, Pat was hysterical. She was the mother of two children, ages ten and thirteen, who suddenly had no father. Even in her raw grief, just hours after receiving the news, a Washington bureaucrat had informed her that despite

Ed's wishes to be buried at West Point, the three fallen astronauts would all be laid to rest at Arlington National Cemetery.

"Give me the guy's name," Borman said.

He had the man on the phone a minute later.

"It's already been decided in Washington," the man insisted.

"I don't give a good goddamn what's been decided," Borman said. "Ed wanted to be buried at West Point and that's what's going to happen, and I'll go all the way to President Johnson to make sure it happens, so you better fucking well do it."

Four days later, White was buried at West Point. Borman and Lovell were among the pallbearers. Anders also attended.

After the funeral, Borman began his work on the investigative committee convened by NASA. He was the only astronaut on the panel, a sign that NASA considered him to be among its best. His first job was to help supervise the disassembly of the Apollo 1 spacecraft at Cape Kennedy in order to determine the cause of the fire. Days later, he became the first astronaut to enter the cabin. He found a burned-out nightmare. Rows of equipment and panels had been charred and covered in soot, debris was scattered everywhere. Hoses connecting the astronauts to their life support systems were melted. No matter where he looked, Borman could see no color, only grays and blacks.

That night, he joined Slayton and others at a restaurant in Cocoa Beach called The Mousetrap, a NASA haunt. Borman seldom drank to excess, but the smell of the scorched spacecraft needed bleaching, and he started in early. He raised toasts to his fallen brothers, then threw his glass into the fireplace. White was among the straightest arrows Borman had ever known—honest to a fault, a true patriot, and a man who didn't mess around with the sports cars or fast women so readily available to astronauts. For both men, family came first. The Bormans and Whites often shared a house on a lake near Houston for fishing trips. Borman couldn't remember missing someone as much as he missed Ed White that night.

Borman spent the next two months inside the burned spacecraft, studying the design, searching for flaws, making fixes in his mind. In April 1967, Congress held hearings into the cause of the fire, and Borman was called to testify.

Much of the questioning was aggressive and antagonistic, full of second-guesses and should-haves and pointed fingers, but Borman held firm, hiding nothing and acknowledging NASA's responsibility, but never allowing congressmen to kick the agency just because it was down. He still ached for the loss of his friend, Ed White, but never allowed those emotions to spill into his report. Near the end of the hearings, he offered some of its most memorable testimony.

"We are trying to tell you that we are confident in our management, and in our engineering, and in ourselves," Borman said. "I think the question is really: Are you confident in us?" A few days later, he told lawmakers, "Let's stop the witch hunt and get on with it." At NASA, it seemed there wasn't a person, from the administrator to the janitors, who didn't cheer him on. In the end, Congress took his advice and NASA continued on its mission to land men on the Moon.

Having survived the inquest, NASA approached Borman with an extraordinary offer: Take temporary leave from the astronaut program to head up the team tasked with implementing design changes to the command module. He accepted on the spot. He and others worked to make the new version of the capsule the most advanced, and safest, spacecraft ever built.

Borman could only hope there hadn't been another tragedy as he landed his jet at Ellington Air Force Base and made his way to Slayton's office. He suspected something unusual was afoot when he was asked to close the door behind him. Slayton addressed him without even sitting down:

"We just got word from the CIA that the Russians are planning a lunar fly-by before the end of the year. We want to change Apollo 8 from an Earth orbital to a lunar orbital flight. A lot has to come together. And Apollo 7 has to be perfect. But if it happens, Frank, do you want to go to the Moon?"

The idea startled Borman. Apollo 8 was meant to fly in December, just four months from now, but certainly not to the Moon. Apollo 8 was a conservative mission designed for low Earth orbit, perhaps at 125 miles altitude. It was one of several essential steps leading up to a manned lunar landing, hopefully before the end of 1969. Everything went in steps at NASA. Everything.

But Slayton meant exactly what he said. He wanted Borman to change missions and fly to the Moon. At a distance of 240,000 miles. In just sixteen weeks. Slayton didn't discuss the fact that the lunar module couldn't possibly be ready by then. He didn't discuss any of the other myriad reasons NASA couldn't be ready to fly men to the Moon by year's end. In fact, Slayton gave very few additional details. He didn't even ask if Borman cared to talk things over with his wife or crew.

Borman would have been justified in taking days, if not weeks, to consider such a proposition. And yet Slayton needed an answer, and he needed it now. Borman understood the urgency. If the Soviet Union sent men to the Moon first—even if those men didn't land—it would score a major victory in the Space Race and deal a devastating blow in the Cold War between the United States and Soviet Union. The mission Slayton was proposing would be exquisitely dangerous. But it also had the power to change history. Now, suddenly, it all depended on the decision of Frank Borman and his crew.

Chapter Two

●

THE SPACE RACE

ON THE MORNING OF SATURDAY, OCTOBER 5, 1957, THE WORLD awoke to headlines announcing that the Soviet Union had launched the world's first satellite. The shiny silver ball, a little more than twice the size of a basketball, was called Sputnik, Russian for "satellite" or "fellow traveler." It was launched by a rocket from the Baikonur Cosmodrome in Kazakhstan and orbited Earth every ninety-six minutes at altitudes between about 140 and 590 miles. Never before had human beings managed to hurl an object out of Earth's atmosphere with such speed that it became part of the cosmic realm. It hardly seemed real. Man had made his own moon.

At first, Americans marveled at the accomplishment, and the best part was that they could witness it for themselves. The Soviets provided radio frequencies on which Sputnik broadcast a beep every three-tenths of a second, along with the satellite's overhead location. Anyone with a shortwave radio could listen to Sputnik. Anyone with a pair of binoculars (or good eyes) could see it, or more likely its carrier rocket, streaking overhead. Millions of Americans gathered outside or by their radios to take in this flash from the future.

But as Monday came, America's weekend of wonderment gave way to darker realities.

The United States was the most technologically advanced nation in the world; twelve years earlier, it had helped end World War II in dramatic fashion when it used the nuclear bomb it developed in strikes against Japan. It should have been the first to put a satellite into orbit. Instead, on the same night that Sputnik launched, CBS aired the debut episode of *Leave It to Beaver*, a sitcom about a squeaky-clean family living in picket-fenced suburbia with all the modern conveniences. To many, it seemed America had been caught fat and happy—becoming Cleavers— while the Soviets had leaped ahead.

And who were the Soviets, anyway? To most Americans, they comprised a technologically backward people living in an all-gray country with a peasant economy and prewar tractors. Yet overnight, they'd made one of history's great scientific breakthroughs. That changed the balance of power; anyone could see it.

"If the Russians can deliver a 184-pound 'Moon' into a predetermined pattern 560 miles out in space," wrote the *Chicago Daily News* that Monday, "the day is not far distant when they could deliver a death-dealing warhead onto a predetermined target almost anywhere on the Earth's surface."

Stories like that whipped the nation into a frenzy. The pitch increased on Tuesday when it was learned that the Soviets had detonated a newly designed hydrogen bomb, one more powerful than any they'd ever tested. Already frightened, many Americans flew into a panic.

Five days after Sputnik's launch, President Dwight Eisenhower, the legendary general and hero of World War II, gave a press conference in which he seemed uncharacteristically out of

his depth when asked about the Soviet satellite. He spoke halt-
ingly, sounding little like the man who, five years earlier, had
said, "Neither a wise man nor a brave man lies down on the
tracks of history to wait for the train of the future to run over
him." Texas senator Lyndon Baines Johnson was more direct
about the threat posed by Sputnik. Soon, he said, "the Russians
will be dropping bombs on us from space, like kids dropping
rocks onto cars from freeway overpasses." The nuclear physicist
Edward Teller, considered to be the father of the hydrogen
bomb, said on television that the United States had "lost a battle
more important and greater than Pearl Harbor." His warning
was echoed by other experts as Sputnik continued to orbit overhead, passing over American airspace, impervious to gravity and
democracy and all the fears of the greatest nation on Earth.

The United States and the Soviet Union had been allies during
World War II, but their cooperation began to collapse after the
United States dropped atomic bombs on Hiroshima and Naga-
saki in Japan in 1945. The bomb was America's effort to end the
war in the Pacific Theater, but the Soviet dictator Joseph Stalin
saw it as more than that: To him, it was a sign of America's intention to dominate the world. Just fourteen days after Hiroshima,
Stalin issued a secret decree ordering the urgent development of
Russia's own nuclear weapon.

The idea seemed a pipe dream. Twenty-seven million Soviet
citizens had died in the war, and the nation's industries had been
decimated. Cities and villages lay in ruin. People were left home-
less, and food was scarce. An atomic bomb required cutting-
edge technology and the marshaling of vast resources and great
scientific minds. The Soviets could hardly build a good car.

But the Soviet Union still had the biggest army in the world.
And it had proved itself able to sustain massive casualties in war.
So American diplomats paid attention in 1946, when Stalin
blamed World War II on capitalism and promised that the Soviet

Union would overtake the West in science and technology. By now it was clear that good science made good weapons.

This was a new kind of conflict, one that would be fought not with bodies on a battlefield, but with propaganda and threats, military buildups, and the formation of alliances—a cold war. Perhaps most important, it would be a race to see which side could harness technology to achieve things that, until now, had seemed unimaginable.

In August 1949, the Soviet Union successfully tested its first atomic bomb—three years sooner than American experts had believed possible. Memories of bodies burned at Hiroshima and piled at Auschwitz remained fresh in the American psyche. No one had to imagine what a mass annihilation looked like, or to wonder whether human beings were capable of inflicting it on each other—they remembered it all too well.

It was around this time that Americans learned to protect themselves—or at least try to survive—during a nuclear attack. In 1952, in schools across the country, a film featuring Bert the Turtle showed children how to "duck and cover" when they "saw the flash." "We all know the atomic bomb is very danger-ous," the friendly narrator said over footage of children hiding under their desks. "Since it may be used against us, we must get ready for it." By 1954, atomic bomb drills were being run throughout the country.

Most people in the mid-1950s expected nuclear bombs to be delivered by airplanes like the B-29 Superfortress that had dropped atomic bombs on Japan, or the new B-52 Stratofor-tress. But these planes suffered the same vulnerability as World War I biplanes: They could be shot down by the enemy. A better delivery system was needed. And both the American and Soviet militaries knew what it was.

The rocket.

It had been used first in combat by the Nazis, when they fired their V-2 rockets at London and other targets in September 1944. The V-2 had a range of only two hundred miles and was too little too late to change the direction of Hitler's war. The technology, however, was full of potential. Ten years later, both the United States and the Soviet Union were working on missiles that could traverse oceans.

Now, one of those missiles had delivered Sputnik into orbit. America knew it had to answer, and fast, by getting its own rocket and satellite to the launchpad. A space race had begun.

Less than a month after Sputnik, the Soviets launched another satellite, only this time it carried a passenger—a dog known to the world as Laika (the Russian word for "barker"). An eleven-pound Samoyed mix, Laika won hearts the world over as she circled the globe. But Laika was no publicity stunt; she was the first step toward sending a man into space, there was no other reason to do it. But there was every reason to try.

A country that could fly men into space was on its way to learning to migrate them off Earth, colonize the solar system, and station soldiers in space. If putting a satellite into orbit gave a nation an advantage on Earth, the ability to populate outer space with citizens and armies gave a nation an advantage in the universe.

And there was another reason to send human beings into space. If man could leave Earth's atmosphere, he could reach the Moon. Forever it had hung there, beautiful and mysterious, calling to man yet always beyond his grasp. The Moon controlled tides, guided the lost, lit harvests, inspired poets and lovers, spoke to children. The nation that first sent a man to the Moon would have done more than make a giant leap in science and technology; it would have fulfilled a longing that seemed to originate not just in the mind but in the soul.

A few days after Laika was launched, it became apparent that the Soviets hadn't designed the satellite to return safely to Earth.

Western impressions of Communist cold-heartedness only wors-ened as the world waited for Laika to die.

Embarrassed again by a Soviet satellite, the United States pushed to launch its own. On December 6, 1957, two months after Sputnik, a Vanguard rocket, carrying its grapefruit-sized satellite, counted down on the launchpad at Cape Canaveral in Florida. Unlike the Soviets, who conducted space operations in secret, the United States was broadcasting this launch to the entire country on live television.

On ignition, the Vanguard's liquid-fueled engine spat orange flames and the rocket began to rise, but just a few feet up it hesitated, tilted slightly, then sank back to the pad, incinerating in a huge explosion. About all that remained of Vanguard in the aftermath was its tiny spherical satellite, somehow thrown free from the blast and lying nearby, beeping like it had made it into orbit.

The humiliation began even before the cinders had cooled. Media around the world called the project "Flopnik," "Kaput-nik," and "Stayputnik," while the Soviets took the chance to revel in America's embarrassment, offering the Americans a help-ing hand through a United Nations program designed to pro-vide technological assistance to primitive countries.

On January 31, 1958, the United States tried again. This time, the rocket climbed straight up, its whiplash of flames light-ing the midnight sky, witnesses yelling "Go, baby!" as the fire grew distant and the sounds fainter. In a few minutes a 30-pound satellite called Explorer was in orbit around Earth. This was a warning shot that announced how quickly things could change when a country believed its survival to be at stake.

A week later, President Eisenhower, the old general, waged his own battle on behalf of the Space Race. He created the Ad-vanced Research Projects Agency, called ARPA, an innovation center for the military where researchers pushed the boundaries of science and technology. (In the 1960s, the agency would at-tempt to network computers across the United States, a project that became the Internet. In 1972, the agency would add the word "Defense" to its title and be renamed DARPA.) In September 1958, Eisenhower signed into law the National Defense Education Act, which provided billions of dollars for the education of young Americans in science and related subjects. And in October, he opened a space agency, the National Aeronautics and Space Administration, known as NASA, which took on the eight thousand workers and $100 million budget of its predecessor agency, the National Advisory Committee for Aeronautics (NACA). Many of its employees were young scientists, engineers, and visionaries.

In December 1958, just about a year after Sputnik had launched, NASA announced Project Mercury, a program designed to put a human being into orbit around Earth and return him and the spacecraft safely. Seven brave men were chosen for the task from a pool of military test pilots. They would be known as astronauts—"star sailors"—and would explore the oceans of space.

America elected John Fitzgerald Kennedy president in November 1960. He'd accused Republicans of being weak on defense and Communism, and Eisenhower of allowing the United States to fall behind in production of intercontinental ballistic missiles—a so-called missile gap. According to the nation's new president, America could not afford to be second to the Russians in anything.

On April 12, 1961, less than three months after Kennedy's inauguration, tracking stations controlled by American intelligence picked up the flight of a Soviet spaceship and detected something startling inside. Minutes later, the Soviet government announced that they'd put the first man into space—whom they called a cosmonaut, or "universe sailor." And he'd already made a complete orbit around Earth.

For the first time, a man had broken the bonds of his home planet. Yet as the twenty-seven-year-old cosmonaut Yuri Gagarin

whirled around the globe, few knew the extent to which the Soviets had rushed the mission, the myriad risks they'd taken, or
the critical tests they'd skipped.

Near the 108-minute flight's end, after reentering the atmosphere, Gagarin's spaceship began spinning uncontrollably and
plummeted toward Earth. He managed to eject and parachute
down, unharmed but almost two hundred miles off course. He
landed in a field near the tiny village of Smelovka, east of the
Volga River in southern Russia, where he was discovered by a
woman and a little girl. The girl ran away, startled by the sight of
this alien being who had dropped from the sky, but Gagarin
waved his arms and called out, "I'm one of yours, a Soviet, don't
be afraid." He struggled to walk in his space suit but managed to
reach the girl and reveal an incredible truth—he had just come
from outer space.

In 1945, the Soviet Union had lain in ruins. Now, sixteen
years later, it had put the first man into orbit around Earth.
Gagarin was given a parade in Red Square, an event as big as or
bigger than the one held to celebrate the end of World War II.
People cried in the streets and hung pictures of the cosmonaut
in their homes.

Gagarin's flight dealt an even bigger blow to the United States
than did Sputnik. "We are behind," Kennedy admitted at a press
conference. Soviet propaganda rained down from Moscow ex-
tolling the virtues of Communism and the superiority of Soviet
science and technology, and it was hard to argue with any of it—
the Soviets continued to do everything first, and biggest, in
space. And that meant that no matter what Khrushchev claimed
about wanting peace, the Soviets were building their advantage
in war.

Kennedy needed to strike back. He asked his vice president,
Lyndon Johnson, to find a long-term challenge that NASA
might undertake, one that would allow sufficient time for the

space agency to catch up to the Soviets, but one that was so difficult, and so spectacular, it could put America ahead in space for good. Kennedy needed something epic, and he needed to announce it soon.

Just days after Gagarin returned from his trip, a group of about fifteen hundred Cuban exiles trained and armed by the Central Intelligence Agency launched a failed invasion of Soviet-backed socialist Cuba at the Bay of Pigs. Kennedy, who'd approved the mission and then withdrawn his support, was devastated by the failure, knowing the damage it would cause to his reputation and that of the United States. "All my life I've known better than to depend on the experts," he told Theodore Sorensen, his adviser and speechwriter. "How could I have been so stupid, to let them go ahead?" To Khrushchev, the answer was simple: The young American president was indecisive and weak.

Three weeks after Gagarin journeyed around the globe, Alan Shepard flew the inaugural Mercury mission. The former Navy test pilot became the first American in space. The fifteen-minute solo flight inspired ticker tape parades, but facts couldn't be ignored: The astronaut had simply gone up and down, while the cosmonaut had made it into orbit—a significant difference in terms of the technology required. As always, the Soviets were far ahead, and with each victory they made a statement to the world, not just about the superiority of their political system and way of life, but about the future.

On May 25, 1961, just a month after the Bay of Pigs fiasco, Kennedy addressed a special joint session of Congress on "urgent national needs." He warned that a battle was being waged around the world between freedom and tyranny, one in which achievement in space could prove decisive.

Then he threw down a gauntlet.

"I believe that this nation should commit itself to achieving the goal, before this decade is out, of landing a man on the Moon and returning him safely to the Earth. No single space project in this period will be more impressive to mankind, or more important for the long-range exploration of space; and none will be so difficult or expensive to accomplish."

The room stood silent. The United States hadn't even put a man into orbit around Earth; now the president was committing the country to landing astronauts on the Moon, and on an eight-and-a-half year deadline, no less. Even if NASA knew how to fly a man to the Moon—and it did not—it lacked the infrastructure, industry, manpower, and technology required to do it. And yet the president stood there insisting it would be done. And soon.

The stakes could hardly have been higher. If America fell short, its failure could not be denied or buried. It would be proof that the nation couldn't do what its leader said was most important, that its greatest minds had failed, that it might not be the world's best hope for the future. It would weaken morale at NASA. And it would embolden the Soviet Union, a nation that wouldn't hesitate to exploit an American embarrassment for propaganda, or press a military advantage.

And yet . . .

If NASA could meet Kennedy's deadline, it would be a statement—to the American people, the Soviets, and the world—that there was nothing the United States could not do if pushed hard enough, that even after losing round after round in the Space Race, falling behind in missiles and bombs, and suffering a humiliation like the Bay of Pigs, the United States could rise in a way no other nation could rise and pull off a miracle. And that's what Congress seemed to hear as Kennedy kept talking and their applause began to build: that landing a man on the Moon and bringing him back safely might be the single greatest scientific and technological challenge mankind had ever faced, but doing it by the end of the decade was impossible, and it was only by attempting something impossible that a nation could truly know who it was.

While Americans buzzed about Kennedy's plan, the Soviet Union yawned. It remained far ahead in the Space Race, and had even sent a probe 42.5 million miles away, which had passed by Venus a few days before Kennedy's speech. In June, Khrushchev bullied Kennedy during a two-day summit in Vienna at which the men discussed Communism and democracy and the relationship between the two superpowers. "Worst thing in my life. He savaged me," Kennedy told a *New York Times* writer. "I've got a terrible problem if he thinks I'm inexperienced and have no guts."

Four months later, on October 30, 1961, the Soviets exploded a device known as Tsar Bomba over northern Russia. Packing a force of nearly four thousand Hiroshima bombs, it was by far the most powerful nuclear weapon ever detonated or even built; for the briefest moment, it equaled 1.4 percent of the power output of the Sun. The device's blast wave orbited the globe three times and its mushroom cloud rose to more than seven times the height of Mount Everest. The ground around the blast site melted and turned to glass, while people fifty miles away were knocked flat.

A year after Tsar Bomba, Khrushchev placed nuclear missiles in Cuba. Kennedy demanded they be removed. Khrushchev refused, but in October 1962, he was facing a different kind of president. Kennedy ordered a naval blockade of the island. For thirteen days, the United States and the Soviet Union stood on the brink of nuclear war. But Kennedy refused to call off the blockade. Just as it seemed both sides had no choice but to use their nuclear weapons, Khrushchev backed down and removed the missiles. The Cuban Missile Crisis had been among the most tense and dangerous events in American history, but when it ended, the world had a different opinion about the will of John Fitzgerald Kennedy.

In mid-November 1963, Kennedy visited Cape Canaveral, where he was briefed on America's developing colossus, the Saturn V, the 36-story-tall three-stage booster being built to take Americans to the Moon. Standing outside with rocket designer Wernher von Braun, Kennedy shook his head in wonder at it all. These men in shirtsleeves and ties were building machines to take human beings to new worlds.

Six days later, the president was dead from an assassin's bullet.

In the wake of Kennedy's killing, some wondered whether the nation's will to land a man on the Moon might have died with him. The new president, Lyndon Johnson, supported the space program and pushed to keep Kennedy's deadline, but problems with logistics, spacecraft, rockets, and engineering bogged down the American effort. Some NASA analysts put the chances of landing a man on the Moon by the end of the decade at just one in ten. In 1964, the Soviet Union only widened its lead in the race to the Moon.

But NASA wouldn't give up. Over the next three years, the Americans and Soviets volleyed for supremacy in space. Project Gemini, designed to perfect techniques the Apollo flights would use to land men on the Moon, opened a floodgate of progress. In the Soviet Union, the skies darkened. Its space program had managed a few interesting missions, but nothing close to the game changers that had put them so far ahead for so long. By the end of 1966, the Soviets were panicked. For the first time since the Space Race began, they were losing.

The American advantage never looked stronger than on January 27, 1967, when three astronauts rode an elevator to the top of a Saturn IB booster at Cape Kennedy in Florida and strapped themselves into their capsule for a simulated countdown. In three weeks they would do it for real, taking Apollo 1—the kickoff of NASA's new Apollo program—into orbit around Earth.

At 6:31 P.M., one of the astronauts screamed into his microphone a word that sounded like "Fire!" Two seconds later, another cried out. His first word was unclear—either "I" or "We"

—but the rest was unmistakable: "got a fire in the cockpit!" That was followed by garbled, desperate words and an agonized scream. Some thought they heard an astronaut saying "We're burning up!"

After that, there was nothing but silence.

Flames spread through the capsule. None of the astronauts could overcome the cabin's highly pressurized atmosphere and move the inward-opening hatch. Seconds later, the capsule ruptured. Technicians rushed to the scene but were beaten back by heat and fire; almost six minutes passed before they could get inside. Rescue personnel found the crew, already expired from asphyxiation, their space suits fused to the melted interior of the spacecraft. Seven hours passed before the bodies could be removed.

Until now, the American space program had owned an excellent safety record; even a chimpanzee named Ham, who'd flown on a suborbital mission in 1961, had come through it safely. Suddenly, three American heroes had died without ever leaving the launchpad, and in a way that seemed entirely preventable. Hundreds of grown men at NASA were reduced to tears by the accident.

Media reports blamed an electrical spark for igniting the pure oxygen environment of the spacecraft's cabin. But to many, there seemed a more basic explanation. "There's reason to believe that establishing a deadline of 1970 for the Moon flight contributed to their deaths," said NBC News anchor Frank McGee. Like many, he thought that by rushing, NASA was risking safety.

After surviving the congressional investigation into the fire, and enduring months of delay while instituting new safety measures, NASA was ready to resume flight operations. On November 9, 1967, controllers counted down the final seconds to the launch of Apollo 4 (Apollo 2 and 3 had been canceled in a reorganization after the fire). This would be the first test of the massive Saturn V booster, a rocket that was orders of magnitude more powerful than any NASA had ever launched, and the only one capable of taking a man to the Moon. The agency dared not put a man on board.

At 7 A.M., the rocket's five enormous engines ignited, sending shock waves of sound and light and energy in every direction as 7.5 million pounds of thrust lifted the six-million-pound behemoth up and away from the launchpad. Three miles away, plaster dust fell from the ceiling in the Launch Control Center, while windows shook at the Howard Johnson's Motel twelve miles from the launch site. Describing the event for a live television audience, CBS anchor Walter Cronkite grabbed the plate glass window of his booth to keep it from collapsing.

"Our building is shaking here!" Cronkite said with uncharacteristic exuberance. "Oh, it's terrific! The building's shaking! This big blast window is shaking and we're holding it with our hands! Look at that rocket go into the clouds at three thousand feet! The roar is terrific! Look at it going!"

The flight worked, every part of it, almost perfectly. It was clear now that America stood a fighting chance, not just of putting a man on the Moon, but of doing it by a long-dead president's impossible deadline.

NASA kicked off 1968 by flying Apollo 5 in January, an unmanned test of the lunar module, the landing craft that would shuttle astronauts between the orbiting spacecraft and the lunar surface. The mission used a smaller rocket, and despite a few problems it was classified a success.

And then came Apollo 6.

It would be just the second test of the Saturn V, a necessary step before NASA would certify the booster for manned flight. Lift-off was proceeding normally on the morning of April 4, 1968, but just a few minutes into the flight, things started to go wrong. The rocket's first stage began to "pogo"—to shake violently up and down. Pieces of the spacecraft flew off. Later in the flight, two of the five engines on the second stage shut down prematurely. Still, the third stage struggled into orbit, but its engine—the one required to send Apollo to the Moon—failed to reignite. A backup plan was put into effect, but the reentry of the command module into Earth's atmosphere was too slow to fully test the heat shield.

To many at NASA, the ten-hour flight had been a disaster. By the time the Apollo command module splashed down into the ocean, any chance for a lunar landing by the end of 1969 looked to have burned away.

"What was illustrated," wrote *The New York Times*, ". . . was the extraordinary difficulty of assuring that every one of the literally millions of components in such an extremely complicated system as the Saturn 5 works perfectly. . . . This fact argues for a slow but sure approach to future Apollo tests, rather than an adventuresome policy aimed primarily at completing the job by the end of 1969."

On the same day that Apollo 6 went haywire, United States intelligence agencies delivered a report on the Soviet space program. It was marked TOP SECRET and went only to high-ranking government policymakers and top NASA officials. It read:

> The Soviets will probably attempt a manned circumlu-
> nar flight both as a preliminary to a manned lunar
> landing and as an attempt to lessen the psychological
> impact of the Apollo program.

That much wasn't news. But the estimate on *when* it would happen jumped off the page. The report said that 1969 was more likely for this manned circumlunar flight. But the second half of 1968 was entirely possible.

NASA had no plans to send men to the Moon in 1968. The

soonest they'd be ready to try was mid-1969, when Apollo 10 would orbit the Moon—a test run before Apollo 11 attempted a landing.

By that time, a cosmonaut might already have reached the Moon. And that would be more than just the greatest technological achievement in history. It would be a definitive victory for the Soviets in the Space Race. The landing would still matter, of course. But no one ever again would ask, "Can we get there?" By that time, someone else would have answered, "We did."

NASA had little choice but to keep working. But as spring turned to summer, there was more bad news for the agency. Plagued by design and production problems, the lunar module had fallen behind schedule. Engineers reported that a fix could take six months or more. That threatened to delay several planned Apollo flights—including those to the Moon.

By early August 1968, things looked dire for the American space program. The Saturn V rocket was in no shape to fly with a crew aboard. The Soviets looked ready to send men around the Moon by year's end. And now, because of issues with the lunar module, Kennedy's end-of-decade deadline for a lunar landing was slipping away.

NASA always proceeded deliberately and carefully. They didn't skip ahead; the risks of manned spaceflight were simply too great. But their hand had been forced. So Deke Slayton had to ask Frank Borman an unthinkable question: Will you and your crew go suddenly—in just four months' time—to the Moon?

Now Slayton needed an answer.

Chapter Three

●

A SECRET PLAN

BORMAN HAD NO IDEA WHAT SLAYTON'S PROPOSED MISSION entailed. He did know, however, that NASA couldn't be ready to go to the Moon in just four months. He knew the agency had yet to build essential systems, calculate proper trajectories, solve problems with its Moon rocket, determine fundamental navigation, develop software, even make a basic flight plan. And he knew how badly the lunar module had fallen behind schedule.

Borman hadn't joined NASA for the usual reasons. He had little interest in exploration, adventure, or pioneering. He didn't thrive on speed or adrenaline. Even the glamorous perks of the job—the availability of beautiful women, discounts on Corvettes, the public's adoration—meant nothing to him. He'd joined NASA for a single purpose: to fight the Soviet Union on the world's new battlefield, outer space.

Before Slayton's question could settle, Borman gave his answer.

"Yes, Deke. Let's go to the Moon."

Slayton didn't need any more than that. He thanked Borman and warned him to keep the information on a need-to-know basis. A few minutes later, Borman was in his airplane and headed back to his crewmates in California.

Flying always focused Borman's mind, and now, cruising at 600 miles per hour, he began to see what a dangerous business he'd signed up for. He believed his crew to be the best at NASA, but four months might not be enough for even this crew to prepare for a journey to the Moon. He had no idea how the space agency would do its part to be ready by December. He could only trust that NASA had carefully crafted the mission, whatever it was, and had taken their time to work out the science.

In fact, much of the plan to send Apollo 8 to the Moon had been contemplated by George Low on the beach just five days earlier. As for the science—that would require some faith.

To fly to the Moon and land a man on its surface, the Apollo spacecraft required three components:

> Command Module—the cone-shaped spacecraft where the three astronauts lived, worked, and conducted most of their mission

> Service Module—the storehouse for the craft's life support systems, its electrical power, and a large rocket engine with sufficient propellant

> Lunar Module—the small landing craft that shuttled two astronauts between the orbiting spacecraft and the lunar surface

NASA needed to test all three modules—both in Earth orbit and around the Moon—before it could attempt a lunar landing. For months, this is how the test schedule stood:

FLIGHT	OBJECTIVE	LOCATION	ESTIMATED DATE
Apollo 7	Test Command and Service Modules	Low Earth orbit	September/October 1968
Apollo 8	Test Command, Service, and Lunar Modules	Low Earth orbit	December 1968
Apollo 9	Test Command, Service, and Lunar Modules	High Earth orbit	February 1969
Apollo 10	Test Command, Service, and Lunar Modules	Lunar orbit	Mid-1969
Apollo 11	Lunar landing	Lunar surface	Late 1969

Apollo 1 had ended in a fatal fire in early 1967. Apollo 2 and Apollo 3 had been canceled after the fire. Apollo 4, Apollo 5, and Apollo 6 had already flown, each unmanned and in Earth orbit.

Everything changed the morning that Low returned from vacation. Even before getting his coffee, he called his secretary, Judy Wyatt, to his office at the Manned Spacecraft Center in Houston.

"I want you to keep a log of people I talk to," he said. "And I want it kept under a secret cover sheet."

Low made a list of key NASA managers and engineers and asked Wyatt to call them in. She'd worked for her boss for two years and knew him to be a precise and serious man. His subjects and verbs always agreed, even while he gave dictation, and he would sunbathe on the weekends with his briefcase beside him, never missing a chance to think or plan. His comments, added to nearly everything that crossed his desk, were made only in green ink; his felt-tipped markers became known as "green hornets." More than anything, Low was efficient; he did not ask for things or take a person's time more than was necessary. On this day, he was requesting the time of several important men, and right away.

He began by seeing two experts—one on NASA safety rules,

the other on trajectories, the flight path a rocket would take to its destination.

Then he called in a legend.

Christopher Columbus Kraft was already a name known to America when he walked into Low's office that morning. An aeronautical engineer by training, Kraft had begun his career in 1944 at age twenty at NACA, the precursor agency to NASA. Small, fit, and serious, with slicked-back black hair graying at the temples, Kraft had masterminded the concept of Mission Control, a central location where nearly all aspects of a spaceflight were managed and supported. By 1965, he'd appeared on the cover of *Time* magazine. In the accompanying article, he compared his role as flight director to that of a symphony conductor. "The conductor can't play all the instruments—he may not even be able to play any one of them," Kraft said. "But he knows when the first violin should be playing, and he knows when the trumpets should be loud or soft, and when the drummer should be drumming. He mixes all this up and out comes music. That's what we do here." The magazine noted that Kraft took "an almost angry pride in his work"—an assessment with which many at NASA agreed. Now he was the director of flight operations for NASA, responsible for the overall planning, training, and execution of manned spaceflight. Whenever Low had a problem, he went to Kraft. Almost always, that problem got solved.

The men shook hands, then Low put a question to Kraft: Could Apollo 8 fly to the Moon in December?

Kraft could have spent the day listing all the reasons why that was impossible. Instead, he just asked, "How?"

"By leaving the lunar module behind," Low said.

The command and service modules, Low reminded Kraft, were in fine shape. Technically, there was no reason those two components couldn't fly without the troubled lunar module, leapfrog the missions of the next two Apollo flights, and go directly to the Moon.

The idea seemed heresy to Kraft. No man had ever flown

more than 853 miles above Earth's surface. Now Low was proposing to send three astronauts a quarter of a million miles away, and to do it half a year sooner than anyone at NASA had planned. As if that weren't enough, Low was proposing to skip not one but two preparatory Apollo flights, violating one of NASA's foundational philosophies: that missions be incremental to assure mastery and success.

And yet Kraft saw elegance, even genius, in the plan. Low wasn't proposing to land Apollo 8 on the Moon, just to fly around it, so no lunar module was necessary. By going in December, NASA could prove many of the systems and procedures, and much of the equipment and technology, required for a lunar landing. It could gain valuable deep space experience, and avoid the months of downtime that would come from delaying Apollo 8 until the lunar module was ready. That would put the agency back on track to make Kennedy's deadline. And there was another benefit: A December launch gave America a chance to beat the Soviets to the Moon.

Still, the logistical challenges seemed insurmountable to Kraft.

Mission Control would need to be readied, trajectories and navigation calculated, an entire deep space communication network finished, an astronaut crew quickly trained, the flight control team brought up to speed and made confident, new software written, instrumentation calibrated. Even if Apollo 8 somehow flew to the Moon and back, NASA would not, as matters presently stood, be able to retrieve the crew, as the agency had yet to schedule an operation for recovering the astronauts when their capsule splashed down in middle of the ocean. Engineers hadn't even run a trajectory analysis to account for the phases of the Moon in December, or lunar lighting at that time of year, or the position of the Moon relative to Earth during such a flight.

Even if NASA could manage all that, the risks of undertaking a lunar mission in December were enormous. Kraft could hardly scribble a list of them fast enough on his steno pad, but two stood out above the rest.

First, the Saturn V rocket—the only one powerful enough to reach the Moon—had never flown with men aboard. It had been tested only twice, the second time in April, when it had suffered near-catastrophic problems. If Apollo 8 was to go to the Moon in December, there wouldn't be time to test the rocket again. The next time the Saturn V rocket flew, it would be with the crew of Apollo 8 aboard.

Second, the lunar module also served as a backup engine—a lifeboat of sorts. Going without it meant that if Apollo 8's single engine failed or malfunctioned at the Moon, the crew could smash into the lunar surface or be stranded in lunar orbit or fly off toward the Sun.

And yet Kraft couldn't bring himself to say no. He asked for a day to study the problem, then met with several experts. He returned to Low's office the next morning with startling news.

Kraft thought Apollo 8 should do more than just go to the Moon in December. He thought Apollo 8 should *orbit* the Moon.

That nearly knocked Low from his chair.

A lunar fly-by, as Low had proposed (and the Soviets planned), required only that a spacecraft be pointed at the Moon. If aimed precisely, it would be pulled in by the Moon's gravity, whipped around its far side, and slingshotted back toward Earth, all without relying on the complex engine burns and calculations that were required to enter and exit a lunar orbit. That made things simpler by an order of magnitude, because it put gravity in charge, not engineers and rockets. In essence, Low had wanted Apollo 8 to fly a classic figure eight from Earth around the Moon and back again. Engineers had worked on this so-called free return trajectory for years, and NASA was certain it was sound.

Now Kraft was suggesting Apollo 8 do something much more difficult. Entering and exiting lunar orbit—whereby the spacecraft would slow itself enough to become captured by the Moon's gravity, then speed up again to leave—required intervention. Engines had to be fired, altitudes changed, speeds modified, navigation altered, and countless other adjustments made. All of it required complex calculations, software, training, and planning far beyond what was required for a free-return flight, and little of which NASA had in its current arsenal.

Yet the benefits of orbiting the Moon could be immense.

Putting Apollo 8 into lunar orbit would provide NASA with all kinds of experience it needed for the upcoming landing mission. Everything from deep space maneuvers to rocket firings to navigation to communications to propellant consumption to life support systems could be tested under the same conditions NASA would face when landing men on the Moon. New mission rules and procedures could be put through their paces, simulations appraised, training revised. And once the spacecraft arrived, the crew of Apollo 8 could photograph the Moon from up close, scouting potential landing sites for the lucky successors who would be the first to step onto its surface.

Low saw the beauty in Kraft's upgrade to his plan. The two men hurried to the offices of other top managers at NASA's Houston buildings, who quickly agreed that sending Apollo 8 to the Moon in December might be the boldest and riskiest and most important mission NASA ever attempted. Now they needed to know whether their NASA counterparts—in Washington, D.C., at the Marshall Space Flight Center in Huntsville, Alabama, and at the Launch Control Center at Cape Kennedy in Florida—would agree.

On that very same morning, Friday, August 9, Low, Kraft, and other top brass in Houston scheduled a meeting with leaders from all the primary NASA centers. Ordinarily, it would have taken a week or more to get all these men together. On this day, they were given until 2:30 P.M. to pack a sandwich, find an airplane, and get to Huntsville, where the meeting would take place.

Everyone arrived on time.

Gathered around a conference table, the twelve men in the room represented a murderers' row of NASA brass. Among them was Wernher von Braun, the world's most renowned rocket designer and the director of the Marshall Space Flight Center. Von Braun had been a member of the Nazi Party and was instrumental in developing rockets, including the infamous V-2 that Hitler launched against targets in Europe. Von Braun surrendered to the Americans in 1945 and went to work for the United States Army, designing rockets. But it was in 1960 that he was charged with one of the most important tasks in the history of space exploration—developing the Saturn super booster that would take men to the Moon. He became the chief architect of the Saturn V—the most powerful machine ever built—and the only vehicle in the world capable of making George Low's vision for Apollo 8 come true.

Notably absent from this meeting was NASA's administrator and top boss, James E. Webb, who was attending a conference in Vienna, Austria. Given the sacrilege that was about to be discussed, it seemed just as well that Webb was thousands of miles away.

Low and others from Houston made their pitch to send Apollo 8 into lunar orbit on a flight scheduled for December. Spirited discussions broke out, ricocheting from man to man, about the benefits and dangers of flying such an audacious mission, and about how to solve all its unsolvable problems. Finally, it came time to take a poll of the men in the room.

The groups from Houston, Washington, and Cape Kennedy agreed: Apollo 8 would be the most difficult and dangerous mission NASA had ever flown. But with unprecedented effort—and a good dose of luck—it might be done. It was worth the risk, these men thought, to keep Apollo on track. And it escaped the notice of no one that there would be a history-changing bonus to flying in December: If Apollo 8 made the lunar journey, America might beat the Soviet Union to the Moon.

That left the group from Huntsville, and the matter of the rocket. Neither von Braun nor anyone else in the meeting needed to be reminded of the Saturn V's recent problems. And yet the mission was impossible unless the rocket could be made ready.

The Moon hung on von Braun's verdict. He thought for a moment, then spoke.

In terms of distance traveled, von Braun said, the Saturn V did not know or care how far the spacecraft went. Like all thoroughbreds, it was built to be pushed to the limit, and in this case the limit was the Moon. And so he did have a verdict.

"You don't give us much time," he said. "But it's a great idea. We just have to see if we can get everything together. But we will try."

The matter settled, the group agreed to adjourn, but not before making a pact.

First, they would take a few days to study the myriad risks and challenges of changing Apollo 8's mission, smoking out any "showstoppers"—problems that could not be solved in time for a December lunar orbit mission. Any of those, and the new plan for Apollo 8 would be off.

Second, they would not breathe a word of this to anyone. It would be hard enough to convince Webb—not to mention Congress and the president—that rushing to the Moon was a good, or even a sane, idea. If word leaked before they ruled out any seemingly insurmountable roadblocks, Washington was sure to bring down the hammer on the plan before it got started.

They would talk again in five days. If all looked good then, they would go to NASA's boss for the go-ahead.

And that's how things stood a day later, when Slayton got down to choosing a crew. He might have given the new mission to Jim McDivitt, who was currently assigned to command Apollo 8. But McDivitt's crew had more experience with the lunar module than did Borman's, so Slayton decided to keep McDivitt ready for when the troubled module was finally flightworthy. He pitched Borman on the new plan for Apollo 8 the next day.

Now, Borman flew back to tell his crew of their new mission, one that hadn't even been officially approved. He'd answered for them in Slayton's office, never imagining they might say no. Yet this was the most dangerous mission NASA had ever contemplated. Borman assumed they'd be as eager as he was to take a sudden shot at the Moon, but there was every chance he was wrong.

Sometimes Borman used the T-38 to do aerobatics, looping and rolling to help clear the cobwebs after a hard day's work. This time, he flew level and fast, back to his crewmates in California in the straightest line a test pilot ever flew.

Chapter Four

●

ARE YOU OUT OF YOUR MIND?

BACK AT THE ASSEMBLY PLANT IN CALIFORNIA, LATER IN THE DAY on August 10, Borman found his two partners and pulled them aside.

Jim Lovell had joined NASA along with Borman in 1962 as part of the "New Nine," the second group of astronauts enlisted by the agency. Like Borman, he was forty years old and a test pilot, but the similarities seemed to end there. Since boyhood, Lovell had been thrilled by rockets and the idea of space travel (he'd gone so far as to attempt to build a liquid-oxygen-powered booster while in high school), and he remained dazzled by the idea of exploring the cosmos. He was also, by most everyone's account, as warm and friendly a guy as one could meet.

Bill Anders was just thirty-four, five and a half years younger than his two crewmates. He'd come up through the ranks as a fighter pilot, not a test pilot. That alone generated contempt from some of the older astronauts, most of whom were test pilots and didn't fully see the daring in climbing into already proven machines. Perhaps even worse for Anders, he was an intellectual among men more immediate and visceral, a holder of an advanced degree in nuclear engineering, and who the hell needed that on the way to the Moon? Still, when people saw him fly they knew they were watching something special. And while it was true he'd flown airplanes already certified by the likes of his colleagues, others could see that he could turn those birds around and shoot most anyone's ass out of the sky.

Borman gathered his crew outside the test bay where they'd been working on the command module.

"Things have changed," Borman said. "If everything goes right with Apollo 7, they want to send Apollo 8 to the Moon by the end of the year. And we are now Apollo 8."

It took a few moments for Lovell and Anders to process what they were hearing. The Moon? By December? Us?

And there was more, Borman said. Apollo 8 would go without a lunar module. And it would orbit the Moon.

Lovell could not fight back his smile.

Oh, man, this is great! he thought. *This is what I've been dreaming about.*

He could see the genius in the plan right away. And the personal benefits weren't lost on him, either. A lunar mission would spare him another Earth-orbital flight, two of which he'd made already as part of the Gemini program. Best of all, it positioned him to do what he loved most—explore and pioneer—and there seemed no better way to do it than by becoming the first man ever to fly to the Moon.

Anders saw it differently.

This new mission would kneecap his chances for landing on the Moon. He'd trained as a lunar module pilot; unless he messed that up, it meant he would walk on the Moon one day. But this new mission had no lunar module, so his duties would shift to the command module, and guys who flew command modules didn't land on the Moon. Five minutes ago, Anders would have put his chances of walking on the lunar surface at 80 percent. Now they'd slipped to between slim and none.

It was a Saturday and the end of the workweek for all of them. The men packed away their gear and climbed into their T-38s, Borman and Lovell in one, Anders by himself in another, and took off into the clouds.

In the backseat of Borman's plane, Lovell began sketching an image on his kneepad—a big Earth in the foreground, a smaller Moon in the background, with a figure eight drawn around the two bodies—it formed both the mission trajectory and its designated number—eight. "What a natural thing for Apollo 8," he thought, and he knew this would be a fine insignia for the patch the crew would wear on their way—on mankind's way—to the Moon.

It was just a few minutes' drive from Ellington Air Force Base to the astronauts' homes in the suburbs. Each man would have preferred more than the usual one day per week at home with his family, but the rigors of training required subordinating family—and everything else—to the mission at hand. In fact, the men and their wives felt lucky. Borman, Lovell, and Anders had served in the military and knew what it was like to be away from home for long stretches. And each of them had friends fighting thousands of miles away in Vietnam, and they gladly would have served there, or anywhere else the country needed them to fight. The couples had all become expert at making the most of twenty-four hours of family time every week.

None of the men had called home to discuss the proposed new mission with his wife. It hadn't even crossed their minds. It wasn't that they didn't respect their wives' opinions. It wasn't even that they were living in a male-dominated culture. These were military men, and even though NASA had been set up as a civilian organization, it was clear to all astronauts—and to their families—that NASA assignments were orders.

Borman greeted his wife with a kiss and told her about the new mission. As always, Susan smiled and clasped his hand. Inside, she was dying.

The Moon? she thought, trying to absorb what he was telling her. It was August. NASA hadn't even tested the command module yet. December—that was what, four months away? Usually crews trained for a year or more. To the *Moon?*

She told Frank how proud she was, how important the mission sounded, that there was no better man for the job. Then she turned and went into another room, where she wished she could kick down a door. *They're rushing it,* she thought. *They're leap-frogging, they're too anxious to get it going.* Over the course of Frank's career, she'd closed her eyes and hoped for the best, but she could see that this mission was different, that she needed to stop living in a cocoon and pretending her husband would always be home for Sunday dinner, because this time Frank wouldn't just be running another test flight—this time he would be leaving the world.

As always, Frank thanked God for Susan. She always supported him, never made him worry about her or their two teenage boys. He had no inkling of what was going on inside her, or how badly she'd been hurting since Apollo 1 had taken Ed White, the husband of her close friend Pat White. Susan knew Frank had enough pressure at work, and she considered it her mission to make home a place where he never worried.

Borman told the news to his two sons, Fred and Edwin. To the boys, the Moon sounded pretty cool. Borman would have showed them where he was going if only he'd owned a lunar map.

At his home, Anders shared the news with his wife, Valerie. Even as he spoke she thought *This is a big and scary change,* but she also had been steeling herself to danger since she was a little girl (her father had been a motorcycle-riding California Highway Patrolman), and she believed beating the Soviets to be a worthy goal.

Bill had always been straight with Valerie, and it would do no good to sugarcoat things now. He laid out his thinking on the risks. He thought there was a one-third chance of a successful mission, a one-third chance of a failed mission that managed to make it back home, and a one-third chance the crew wouldn't return at all. He hated to worry her, but he knew if she sensed he was bullshitting, she just would have worried more.

Valerie trusted that these odds were accurate, though they were not numbers any young mother of five liked to hear. She thought about other military wives, some of whom had husbands missing in action, and she remembered that her husband, like most astronauts, would be fighting in Vietnam—eagerly—if he weren't training to fly to the Moon. She kissed Bill and told him she believed in NASA, in the mission, and in him.

On Lovell's arrival home, his wife, Marilyn, made a happy announcement: She'd scored several bargains on clothes for the family's upcoming Christmas vacation in Acapulco. It was hard for her to believe it was happening; she couldn't remember if they'd even attempted a vacation since he'd joined NASA six years earlier.

Jim smiled as Marilyn held up her brightly colored beach buys, but she could see his mind wasn't on Mexico. "Are you all right?" she asked. He motioned her into his study.

"I can't go on vacation," he said.

"I can't believe it!" Marilyn replied. "I've already made all these plans for Acapulco!"

"I'm going somewhere else. Somewhere special."

"Where are you going?"

Jim grinned.

"Would you believe, the Moon?"

Looking into Jim's eyes, Marilyn still saw him as the boy she'd met in high school, who talked about stars and planets, and as the first-year Naval Academy midshipman who'd asked her to type up his term paper that predicted men would someday ride rockets into space. She knew that this was what her husband had been seeking all his life.

Lovell called his four children, ages two through fifteen, into the study and told them the news. He did own a lunar map, and he laid it out on his desk to show his kids where America wanted him to go.

Now that Low and other top NASA managers had decided to shoot for the Moon, they needed to pick the best day to go. A primary objective was to replicate a landing mission to the fullest extent possible, in order to provide the agency with relevant experience for when it ran the real thing. That meant, among other things, figuring out the optimal alignment between Earth and Moon so that the flight required no more propellant than necessary, and so that there would be excellent lunar lighting to scout potential landing sites. Only a few days per month lined up like that, so NASA had to choose well.

Management had access over the weekend to the agency's seven giant mainframe computers, which calculated four possible launch windows. Optimal lift-off would be December 20 or 21, with a splashdown six days later in the Pacific Ocean. That gave Apollo 8 its best look at the Moon, and time to make several orbits, each of which would last about two hours. It also allowed for an early morning countdown at Cape Kennedy, which would give NASA plenty of daylight to rescue the crew if something went wrong during launch.

That all sounded fine to Kraft until he realized what that meant: Apollo 8 would be in orbit around the Moon on Christmas. He knew NASA would be accused of selecting the date for effect, but all he could do then would be to tell the truth: The agency hadn't chosen Christmas, nature had.

For the next several days, many of NASA's top managers and engineers stepped up their already intense schedule and worked around the clock to study the viability of Low's plan, looking for any showstoppers and keeping it a secret from the wider organization. A few days later, they were convinced: It would take a near miracle, but every problem could be solved, every challenge could be met. Now it was time to go to NASA's top boss, James E. Webb, for permission.

Some at NASA doubted that Webb would even listen. The Apollo 1 fire had nearly put him and the agency out of business, and it seemed unlikely he'd risk another tragedy. But they had to make their pitch now if the agency was to have any hope of sending Apollo 8 to the Moon by year's end.

The job of seeking official permission fell to Webb's deputy, Thomas Paine, a young, forward-thinking engineer, and to Air Force general Samuel Phillips, director of the Apollo Manned Lunar Landing Program. Phillips at first wanted to do it in person but then thought better of it—a sudden trip to Vienna by high-ranking officials of the American space program might tip off the Soviets that NASA was planning something big. The better idea was to use a secure telephone line and hope for the best.

Paine and Phillips reached Webb at the American embassy in Vienna. They had reason to hope Webb would see the genius of the new plan. Since Kennedy's speech in 1961, Webb had been a champion for Apollo, protecting and advancing the program with Congress, playing by street rules when necessary. So the men laid out their vision for Apollo 8.

"Are you out of your mind?" Webb yelled.

He began to count off the risks of sending Apollo 8 to the Moon in December, only to grow more indignant with each one, and his list didn't seem to end.

"You're putting the agency and the whole program at risk!" Webb finally said.

And it was hard to argue with any of it. Three astronauts had died a horrific death on the launchpad less than two years earlier. Congress would not abide another three dead, especially if it occurred because NASA had hurried.

And Webb added a final point. "If these three men are stranded out there and die in lunar orbit, no one—lovers, poets, no one—will ever look at the Moon the same way again."

No one had considered that. But it was true of Christmas, too. Borman, Lovell, and Anders would be in lunar orbit on December 25. If they died then, Christmas would never be the same in America. Or maybe in all the world. Every year, it would be a tragic reminder of a mission gone horribly wrong.

Webb had little to gain by signing his name to such a risky plan. And though he hadn't announced it, he planned to resign in a few months, ending his seven-year tenure at NASA. No sense in sticking his neck out for a crazy mission he wouldn't even be around to oversee.

And yet, even as he continued screaming into his telephone, he did not say no. Instead, he said he'd think about it. And he promised to get back to the men the next day.

As NASA awaited Webb's verdict on Apollo 8, Soviet cosmonauts trained for their own December lunar mission. It was a treacherous business, and they were taking risks that were normally forbidden, but with the Moon in the balance these were not normal times.

A day later, Webb found a secure phone line in Vienna and called his men in Washington. He still thought the new plan for Apollo 8 was saturated with risk and danger, and that it could ruin NASA, but he could not deny its potential. He gave Paine and Phillips the go-ahead to prepare for a December lunar launch, but warned that he would not sign off on the plan unless and until Apollo 7 orbited Earth and completed its mission objecctives in the fall. That was all the green light these men needed. Even as they hung up they were dialing top NASA brass: The big guy had spoken. Eight was Go for the Moon.

Now NASA needed a flight plan. Ordinarily, that took months to devise, but time was suddenly a luxury of a bygone era. Early in the afternoon on August 18, Borman met with Kraft and some of NASA's top designers, planners, and engineers in Houston. Everyone had come to hammer out a blueprint for Apollo 8's flight to the Moon. No one intended to leave until it was done.

Borman looked around the office. To him, this was the ideal setup: no committees, no memos, no suits from Washington, just top-notch guys led by Kraft—cigar in mouth and as tough under pressure as a fighter pilot—who would make the final decisions on everything.

"Okay, so we're going to the Moon in December," Kraft said. "Now, let's figure out how we might do that."

Everyone started with the same basic questions: How long should the journey last? How many orbits should we make around the Moon? How high above the lunar surface should we fly? What do we most want to accomplish? What should the crew be doing? When do we want to come home?

The planners took on the subject of lunar orbits first, and their desire was simple: They wanted the maximum number the mission could sustain. Borman's reply was equally simple: Forget it. The meeting was five minutes old and already the players were stuck.

Borman understood their position. Every orbit gave NASA more opportunity to gain experience for a future lunar landing. But they had to understand his first priority, which was the safety of his crew. He, Lovell, and Anders would be flying an unproven spacecraft on a mission riskier than any NASA had ever attempted. To him, each additional orbit was another chance for something to go wrong. As far as Borman was concerned, flying once around the Moon would be a historic accomplishment—and more than enough to beat out the Russians.

Borman's position didn't please Kraft's men, who thought it wasteful to take all the risk of flying to the Moon only to leave early once they got there. Kraft jumped in to calm the planners down.

"What's the absolute minimum you can take?" he asked them.

The men thought about it and came back with an answer: twelve. Since each orbit would last about two hours, that gave the crew twenty-four hours around the Moon.

"Ten is better," Borman shot back.

But the men shook their heads. If Apollo 8 flew only ten orbits, it would splash down in the Pacific Ocean before dawn. That meant if the parachutes malfunctioned, no one could see what was happening.

"What the hell does that matter?" Borman said. "If the chute works, great. If it doesn't, we're all dead and it won't make any difference if anyone can see us."

No one could argue with that.

Kraft asked if his men could accept ten orbits. They nodded. Kraft liked it. Ten was a rational, empirical number. And like that, it was ten orbits.

The question then became at what altitude to orbit. Kraft and his men wanted Apollo 8 to fly just 69 miles above the lunar surface, the same altitude at which the command and service modules would operate during a future landing mission. That required almost unimaginable precision, equivalent in scale to throwing a dart at a peach from a distance of 28 feet—and grazing the very top of the fuzz without touching the fruit's skin. If that weren't daunting enough, the Moon would be barreling through space at nearly 2,300 miles per hour. Toss a peach in the air at 28 feet and now hit the top of the fuzz with a dart. That's what these trajectory experts were proposing to do. And soon everyone agreed to do it.

As the men continued to talk, the details of the flight took shape.

Launch would occur in early morning from the Kennedy Space Center on December 21, when the new Moon would be just a sliver in the sky. Borman and crew would orbit Earth for a short time to check out the health of the rocket and spacecraft. If all looked good, they would attempt to relight the Saturn V's third-stage engine—no sure thing, as it had failed on the test flight in April. If it did work, the engine would push Apollo 8 to a speed of 24,200 miles per hour, enough to break free of Earth's gravitational pull. To date, no human had ventured more than 853 miles away from Mother Earth. Borman, Lovell, and Anders would blast past that in a few minutes. Even then, they would still have to cover nearly 240,000 miles to reach the Moon, about fifty-eight times the distance Columbus had sailed to find his own new world.

The Earth–Moon crossing would last about sixty-six hours. The astronauts would spend much of that time doing navigation sightings, making live television broadcasts, and checking systems. The spacecraft would fly on a precise round-trip pathway, aimed close enough to the Moon's surface to ensure that if anything went wrong along the way, the Moon itself would make the rescue, catching the ship with its gravity and slingshotting it back to Earth, all courtesy of the laws of physics, with no man-made propulsion required.

As Apollo 8 neared its target, the Moon would be moving at more than 2,000 miles per hour, with the spacecraft rapidly accelerating, both approaching nearly the same spot in space. If NASA's figures were accurate, the ship would slide just ahead of the Moon's leading hemisphere, then use lunar gravity to curl behind the lunar far side.

Once the spacecraft went behind the Moon, all communication with Mission Control would be blocked. At that point, if all looked good, Borman would fire the Service Propulsion System, or SPS, engine, which would slow the ship enough to be captured by the Moon's gravity and enter lunar orbit. There was no backup to the SPS. If it didn't fire, Apollo 8 would whip around the Moon and return to Earth. The real problem would come if the engine fired incorrectly: too short or too weak, and the spacecraft would fly off into eternal space; too long or too strong, and it would crash into the Moon in less than an hour.

If it all worked, however, Apollo 8 would enter an irregular orbit, about 69 miles above the lunar surface on the far side and about 200 miles over the near side. For two revolutions (about four hours), the crew would prepare their cameras and observe landmarks. Then they would get ready to fire the SPS engine again, this time to circularize their orbit at a constant 69 miles above the lunar surface. It would then be Christmas Eve morning back in America.

Once in a circular orbit, the crew would do the bulk of its work. For eight revolutions over the next sixteen hours, they would scout candidate landing sites for future missions, take photographs, analyze lighting conditions, and study the effects of gravitational anomalies on the spacecraft's orbit. All the while, Mission Control would be tracking the spacecraft by radio and communicating with the astronauts, except when Apollo 8 was over the far side of the Moon.

And there would be two more television broadcasts. As Borman did the math, he could see that these would come on Christmas Eve and Christmas Day.

During the final two revolutions, the astronauts would get ready to fire the SPS engine again, this time to gain enough speed to get the spacecraft out of lunar orbit and on its way back to Earth. As before, the firing would be done over the far side of the Moon, out of contact with Houston and the rest of the world. It was another critical maneuver: If it misfired, the ship could crash into the Moon or fly off into the void. If it failed to fire, Apollo 8 would become a possession of the Moon. Forever.

But if all went according to plan, the spacecraft would escape lunar orbit and begin its fifty-seven-hour journey home. In the flight's last minutes, the service module containing the engine would be jettisoned, leaving the astronauts in the cone-shaped command module they would ride the rest of the way to Earth. A short time later, the capsule would begin reentry into Earth's atmosphere at near 25,000 miles per hour. No human had ever attempted such a thing, and it had to be virtually perfect or Borman and his crew wouldn't survive.

Angle of attack was everything. Apollo 8 needed to enter a corridor that spanned just two degrees. That was equivalent to finding exactly the right ridge on a coin that had 180 ridges grooved into it. (By comparison, a United States quarter dollar has 119 ridges.) If the spacecraft came in too shallow, it would skip off the atmosphere like a stone on water, going into a large elliptical trajectory around Earth without enough oxygen or electricity on board to get back for another attempt at reentry. If it came in too steeply, it would grind so hard against the atmosphere that the resulting heat and deceleration would burn up and tear apart the capsule. But if it came in just right, the atmosphere would slow the capsule down enough to allow it to survive reentry into the atmosphere and plunge toward Earth. Two degrees—anything on either side of that and the crew was dead.

If Apollo 8 survived reentry and if its heat shield succeeded in preventing its incineration in temperatures that would reach 5,000 degrees Fahrenheit—half that of the surface of the Sun—the triple canopy of parachutes would deploy and the capsule would splash down in the Pacific Ocean about forty-five minutes before first light. The astronauts would stay inside until a Navy recovery crew reached them. By that time, Apollo 8's historic voyage would have ended, after a little more than six days.

Borman looked at the other men in the room. Each wore the same expression: *We know this is impossible, but we still think it can work.* He appreciated their commitment and expertise, but he thought they'd planned too much for the crew to do—every hour seemed loaded with tasks, duties, obligations, checks.

"Forget the TV cameras," Borman said. "It's a distraction."

"No way," Kraft said. "This is history, Frank. This belongs to the American people."

"We're here to do a job," Borman said.

"That's part of the job," Kraft answered.

Borman saw no yielding in Kraft's eyes. The cameras would stay.

Borman still objected that the work plan was too crowded,

and Kraft didn't deny it. There was a lot to do, maybe too much, but six days on a moonshot was an eyeblink, given the risks and expenditures required to get there, so they damn well had to get the most out of it. Anything less and none of them would be doing his job.

That made sense to Borman.

And with that, the plan was complete.

The men checked their watches. It was five P.M. In just four hours, they'd designed a mission that would send the first human beings away from their home planet, have them orbit the Moon, then return home. In a year that was shaping up to be among the most fractious in the nation's history, in which its citizens were rippling with anger and its institutions were no longer trusted, something sublime had occurred in this office. Shaking hands, Kraft and Borman had the same thought: *This was a great afternoon. This was America at her best.*

The two men left the building together. As Borman walked past the other astronauts' Corvettes and climbed into his 1955 Ford pickup, Kraft could only admire him. Even during this technical meeting, Borman had been true to form: direct, principled, and bullshit-free, unwilling to look past minor details or compromise around edges. To many, including Kraft, he seemed the ideal astronaut to command the riskiest flight NASA might ever undertake. To those who knew him best, it seemed Borman had arrived at a crossroads, not just in his career but in his life.

Chapter Five

●

FRANK BORMAN

FRANK FREDERICK BORMAN FIRST LEFT EARTH AT AGE FIVE, IN 1933, when his father took him on a trip from their home in Gary, Indiana, to an airfield in Ohio. There, a barnstorming pilot wedged father and son into the front seat of a Waco biplane and flew them over the countryside. Five-year-old Frank could hardly process the freedom of it all—the open cockpit, the wind in his face, nothing between him and the rest of the world as the machine growled and swooped through an endless sky. The pilot asked for five dollars when the airplane finally settled back on Earth, a fortune during the Great Depression, and the greatest bargain Frank could imagine.

Not long after, Frank's family moved to Tucson, Arizona. His father, Edwin Borman, leased a Mobil service station and tried to make a go of it. The Bormans didn't have much—just a rented two-bedroom home and a 1929 Dodge with creaky wooden spokes. As the Depression moved into the 1930s, Edwin's business suffered and he lost his gas station lease. It was then that Frank saw his dad live by the mantra he'd been preaching forever: *Do not quit, stay in there and pitch.* Edwin took a job changing tires at another garage, then found work driving a laundry truck. Frank's mother opened their house to boarders to make extra money.

At school, Frank's teachers observed him to be bossy and headstrong, a report that didn't surprise his parents. Since the day Frank could walk, he had moved in straight lines and with shoulders pinned forward, a kid compelled to arrive. Not everyone knew where Frank was going, not even his mom and dad sometimes, but it seemed to them a mistake to label the boy rude or abrupt just for pushing past people and things that slowed him down. They'd always told Frank he could be the best at whatever he chose if he did things the right way, with excellence and integrity, no shortcuts.

Edwin and Frank often sat together at their living room card table building model airplanes, some powered by rubber bands, others by tiny temperamental gas engines that screamed like banshees. Frank learned to take responsibility for his creations. Edwin never stepped in and finished the job for Frank, no matter how many times the engines wouldn't fire—even during model airplane competitions, even while judges were waiting. He just let Frank keep working, keep adjusting, until the Borman plane flew better and farther than all the rest.

By the late 1930s, many kids across America had become fascinated by the idea of space travel. Scientists were developing rocket technologies, and the future that these machines promised exploded in color in popular science and adventure magazines, comic strips, and films. Frank couldn't have cared less. Science fiction bored him. If his friends went to see movies about spaceships, he stayed home and built airplanes, the kind of machines that flew for real.

Frank entered high school in 1943, in the midst of World War II. Schoolwork came easily to him, which left him time for deeper pleasures. One day, he wandered over to nearby Gilpin Airport and told the manager he wanted to fly. The man had no problem with Frank's age—fifteen—but warned that lessons cost nine dollars an hour. Frank knew his parents couldn't afford that, but he did some quick mental math. By combining the salaries from his three current jobs—bag boy at Safeway, gas station jockey, and sweeper at Steinfeld's Department Store—he could put himself into the air.

He signed up and was taken to a hangar where he met his instructor, who was about thirty years old, had trained very few students, and was dressed not like a pilot but in Levi's and a white T-shirt, which was unusual at the time for a woman.

In the 1940s, only about a hundred women worked as flight instructors in the United States. Bobbie Kroll was one of them. Frank hardly noticed her gender, she hardly noticed his age, and at once they were together in the cockpit. Miss Bobbie was an ideal teacher. She would not yell or panic, and she remained calm when Frank banked too hard or struggled to come out of a stall. After just eight hours of dual instruction, she turned Frank loose to solo.

For the next three years, Frank continued taking flying lessons, making good grades, and playing quarterback on his high school football team. But his best night of high school came during senior year at a local dance in Tucson, when he spent the evening moonstruck by a golden-haired sophomore named Susan Bugbee. She'd been voted the most beautiful girl in her class, and Frank, a longtime believer in democracy, thought the voters had gotten it right. He was aching to ask her out, but this young man who stared down thunderstorms in small airplanes couldn't stomach the idea of rejection. Instead, he came up with a plan. A friend of Frank's would call Susan on the phone. Pretending to be Frank, he would ask her for a date. That way, if she said no, Frank wouldn't hear it.

Susan said yes. Frank wished he'd heard it.

The two began dating, and right away Frank sensed he'd met

his soul mate. Susan was bright and quick-witted, warm and fun, and loyal to her friends. Sometimes she wrote "Susan Bugaboo" instead of "Bugbee" in her notebooks. She had a mischievous gleam in her eye, the same as when she'd been in elementary school and pulled the fire alarm during a rainstorm as a prank (the nuns were not happy; Susan's father loved it and smoothed things over with the sisters).

Susan's parents were both college graduates, rare in those days. Her mother was Tucson's first female dental hygienist, her father a surgeon who'd moved to Arizona after losing a lung to tuberculosis. Susan had been very close to her father, who took her on house calls and had her join him on his volunteer work to help the underprivileged. They often went on adventures together: on his days off, he would drive her outside the Tucson city limits to the ends of dusty roads, where they would capture tortoises together (she'd keep them as pets for a while, then release them), and Dr. Bugbee would buy his daughter turquoise jewelry from Native Americans who sold their wares from the backs of old pickup trucks. Susan was never as close to her mother, who seemed to resent her for all the attention people paid to her.

One day, when Susan was thirteen, her father had an asthma attack. His oxygen bottle was empty, so Susan's mother told her to run to Johnson's Drugstore and get a new one. Susan got the pharmacist to drive her home, to save time and in case he could help. But by the time she returned, her father lay dead on the floor.

"You're late," Susan's mother said. "You killed your father."

The words devastated Susan. On the spot, she knew she'd never forget them. But something about that incident steeled Susan's spine. From the day Frank began dating her, he sensed an undergirding of strength in Susan. *This girl*, he thought, *can handle anything*.

As high school drew to a close, Frank needed to decide on a future. He wanted to be a fighter pilot—a perfect way to combine flying and defense of his country. World War II had ended nearly a year earlier, but already tensions were building with the Soviet Union. No less an expert in looming tyranny than Winston Churchill now warned that "an iron curtain" had descended across Europe. Frank believed him.

After scoring high on admissions exams, Frank enrolled at the United States Military Academy at West Point in the fall of 1946. Cadet Borman was all baby face and golden hair compared to his classmates. Many had already attended college, and at least half were veterans of World War II. In early fall, Borman tried out for the plebe (first year) football team. He'd been a star high school quarterback, but at this level he didn't have the necessary arm strength. He joined anyway, as the varsity team's assistant manager, in charge of gathering dirty socks and sweaty jockstraps. It was thrilling for Borman, who got to observe head coach Earl Blaik's legendary intensity and to watch one of the young assistant coaches, Vince Lombardi, develop his own military coaching style.

Borman fell in love with West Point. The rules, the order, the discipline—it all seemed designed to tune out distraction and allow a man to get on with what really mattered. As a kid, he'd already been different from his peers—he went after the things that were important to him, as if he were on a mission. At West Point, nothing mattered but the mission. He pledged himself to the academy's motto—Duty, Honor, Country. It seemed to Borman that a person who believed in anything less wouldn't get where he needed to go.

All the while, Borman and Susan continued dating, if only by U.S. mail. She was still in Tucson, and they were separated by more than two thousand miles. West Point did not allow furloughs for plebes, even for holidays. Fearing he'd receive a breakup letter from Susan, Borman struck first, sending a letter to Susan saying they needed to cool their relationship. It only made sense, in light of their distance, his commitment to West Point, and the focus he'd need to make his new dream, of becoming an Air Force general, come true. Susan knew: She was no longer his mission. The letter broke her heart.

By the end of his third year, Borman ranked near the top of his class. For her part, Susan had enrolled at the University of Pennsylvania's dental hygiene school, following in her mother's footsteps. While there, she was offered a contract with the Ford Modeling Agency in New York, which she declined. During quiet moments, Borman wondered if he'd made the mistake of a lifetime by letting her go.

In the summer of 1949, Borman was one of a select few cadets chosen to tour postwar Germany. For him, the biggest impression came at the Nazi concentration camp at Dachau. There, he saw the firing range and gallows used to execute Jewish prisoners, and the ovens used to cremate them. And he saw families, East German refugees, living in tiny stalls in the barracks, separated from other families only by hanging blankets; these were people who'd chosen to give up everything and flee to the West rather than live under Communist rule. The trip sickened and saddened him, and it reinforced his certainty that America was a force for good in the world, a country that stepped up to help suffering people and defend freedom.

When Borman returned from Germany, he only missed Susan more. She had returned to Tucson after earning her degree, and was chosen over seventy-one other contestants as the city's representative at a Mardi Gras festival in Mexico. The local newspaper showed her draped in a silver-blue mink cape and wearing over a thousand dollars' worth of silver and turquoise. In case Borman had forgotten what he'd lost, the newspaper noted that the selection was based on "beauty, poise, personality, charm, and intellect."

Borman graduated eighth in a class of six hundred seventy at West Point. It was a beautiful ceremony, but all he could see were the swarms of girlfriends and fiancées who'd come to shower love on his classmates. His only comfort came from knowing he'd been among those selected for a coveted spot in

Air Force flight training, and from driving his parents back to Arizona in the new car he'd purchased, a blue Oldsmobile Rocket 88 stretch coupe with a V-8 engine and a bench seat in back.

Borman had sixty days' leave before reporting for flight training at Perrin Air Force Base in Sherman, Texas. On the first of those days back in Tucson, he called Susan and asked her to dinner. They hadn't had a date in ages, and she still had hurt feelings from their breakup of three years ago, but she agreed. He took her to a small Italian restaurant on the outskirts of town. They laughed and talked and connected as if they were still in high school; even the owner could see their chemistry because he kept feeding the jukebox and pressing the love song buttons. Borman didn't waver this time, he did what he'd been wanting to do for years—he asked Susan to be his wife. There was no talk of the challenges of a military life or the risks he'd be taking as a fighter pilot. There was just the question—"Will you marry me?"—and her answer—"I will."

A month later, Frank and Susan were married in a Tucson Episcopal church. After honeymooning at the Grand Canyon and in Las Vegas, the Bormans reported to Perrin Air Force Base, then to Williams Air Force Base in Chandler, Arizona. These were fun and adventurous times for the new couple, even if training was risky. Men died from losing control while pushing the limits in these high-performance jets, but it never occurred to Borman that he'd be hurt. Others had survived the training, and he knew he was better than any of them.

Susan never complained about the dangers of Frank's job, the hours it required, or even their tiny home, a trailer with no airconditioning. Once, after Frank's model airplane flew away from him, Susan spent the next day searching the area for miles, knowing how disappointed he was to have lost it. She didn't find it, but Borman was touched that she didn't want him to worry, even about little things.

Soon Susan was pregnant. A month before the baby was due, in September 1951, Borman was transferred for the second time in eight weeks, this time to Nellis Air Force Base in Las Vegas. He protested, arguing that the move was too much for his eight-months-pregnant wife. A captain reminded him, in various shades of blue, that there was a war going on in Korea. Borman gathered blankets and a pillow and turned the bench seat in the back of the Oldsmobile into a bed, tucked in his pregnant wife, and drove to Las Vegas.

On October 4, 1951, Susan gave birth to a baby boy, Frederick. On the same day, Borman flew two missions—no time off for a brand-new father, such was the urgency of wartime training. The work of a fighter pilot was exceedingly dangerous; six men died over one weekend, all at Borman's base. At home, Susan never allowed her husband to see how these accidents made her shake.

New orders sent Borman to the Philippines, closer to the war in Korea, which was just what a fighting man wanted. Still just twenty-one years old, Susan sold the Oldsmobile for the price of a one-way plane ticket and, with baby on lap, made her way to Manila. Another son, Edwin, was born in a Quonset hut at Clark Field in July 1953. A few months later, Borman's tour in the Philippines ended, and so had the Korean War. The battle he'd signed up to fight had faded away.

Borman spent the next several years logging hours in fighter jets, learning to drop atomic bombs, waiting for his chance to defend America. Always he posted the highest marks, blending rare piloting skills with a fighting instinct and a mission-first tunnel vision. Wherever he went, he considered Susan his secret weapon, a partner, mother, and best friend who arranged their lives so that his only worries were in cockpits.

Not all of it came naturally to Susan. Every boom in the sky, every siren on the base, had to be answered by reminding herself, *It's not going to happen to Frank. He's different. Frank's a better pilot than they are. Frank will always be okay.* After leaving the commissary one day at Moody Air Force Base in Georgia, Susan witnessed a midair collision between two jets. She knew Frank was flying at that time. Both airplanes were two-seaters, but only three parachutes opened in the sky. Frantic, she ran toward the billowing black smoke and tried to climb a fence to reach the field, but she was intercepted by a GI, who ordered her to go home. Susan raced back to her neighborhood and banged on the door of Frank's boss. The man's wife let her in.

"What do I do?" Susan asked.

"What you do is wait," the woman said.

And Susan did, for two and a half hours, until Frank landed and called her. He reacted to news of the fatality as he always did, by thinking *That dumb sonofabitch killed himself; it'll never happen to me because I'm better.* It was a defense mechanism shared by many fighter pilots, and Susan bought into it, too. At least for now.

In 1956, Borman was ordered to earn a graduate degree in aeronautical engineering in order to become an instructor at West Point. He enrolled at Caltech in Pasadena, where he kept up with some of the best students in the world. By 1957, he had his master's degree and was teaching thermodynamics and fluid mechanics at West Point.

He loved being back at the place that had shaped him. If anything, Susan loved it more. Her boys were playing little league baseball and learning to swim, she'd decorated the family's apartment, and Frank was home most nights. For the first time since they'd married, it seemed a stable existence, and one that might last.

A few months later, the Soviet Union launched Sputnik. Borman couldn't imagine a bigger blow to national pride, or a clearer indication that America was losing the Cold War. Already a staunch anticommunist, Borman now believed the United States to be facing an existential threat. From that point forward, his thinking changed. If he could do anything to be part of the fight America needed to bring against the Soviet Union, he would do it. Even if the United States needed him to drop an atomic bomb, he wouldn't have hesitated for a second. He didn't want to kill anyone, let alone innocent civilians, but his faith that his country would always act as a force for good in the world trumped all.

In 1960, Borman applied to and was accepted by the Air Force's exclusive Experimental Flight Test Pilot School at Edwards Air Force Base in California. It was in the skies, he thought, that the fight against the Soviets would be decided; technology would determine how high and how fast.

He began training in a Lockheed F-104 Starfighter, flying at 1,600 miles per hour, more than twice the speed of sound. Much of what he did at Edwards was experimental and untested, making it dangerous in ways one couldn't train for, and in ways that he never discussed with Susan.

Borman graduated first in his class academically and second in flying and won the award for best overall student at Edwards. (He would have been first in flying but for a momentary failure to raise a landing gear, a slipup that would bother him for years.) He then signed on to establish a new program at Edwards, the Aerospace Research Pilot Graduate course, designed to prepare future astronauts to fly. He and four other top pilot-engineers would create a curriculum, making sure it best positioned a man for selection by NASA. It did not escape his notice that as an instructor, NASA might consider him to be among the best candidates of them all.

In March 1961, Borman came to a crossroads. NASA was looking to bring on a second group of astronauts and asked top Navy and Air Force pilots to apply. If he had any interest in going into space, now was the time to strike.

Borman didn't thrill to the idea of riding on rockets or exploring the cosmos or even stepping on the Moon. The instant celebrity conferred on astronauts seemed a distraction to him. And yet only NASA could deliver him onto a new battlefield, where technology and futuristic flying machines could help determine whether democracy or Communism prevailed. With the Cold War growing hotter every day, he could think of no more important place to do his part than on the frontier of space.

He talked to Susan. He told her he had a chance to help America, and to make history, but it would require undertaking a new life and unknown risks. Susan answered as she always had: They were a team and she would support him. A short time later, he submitted his application to NASA, joining more than two hundred other highly qualified hopefuls. He endured exams—physical and psychological—and several rounds of cuts as NASA trimmed its list of finalists to about eighty, then to thirty-two. Finally, in the fall of 1962—eighteen months after he first put his name in the hat—Borman became one of the agency's nine new astronauts, selected from America's best to go where mankind had only dreamed of going.

NASA introduced its second group of astronauts to the public at the University of Houston on September 17, 1962. Soon to be dubbed the New Nine by the press, they included James Lovell and Neil Armstrong. All nine had been test pilots and had studied aeronautical engineering. All were married and had children. From the moment he stood beside these men, Borman could tell he was among a rare group, talented and competitive beyond any he'd met.

The new astronauts became instant celebrities. As with the Original Seven, each received a contract with *Life* magazine and Field Enterprises that paid him $16,000 a year for exclusive access to his and his family's personal stories. For her part, Susan would be obliged to speak at luncheons and urge young mothers to buy *World Book* encyclopedias (published by Field Enterprises) for their families.

NASA assigned each new astronaut to a specialty. Borman's was boosters, the rockets that lifted spacecraft off Earth and into orbit and beyond. His focus would be on a crucial aspect—the crew safety and escape systems. Borman and his colleagues would spend hundreds of hours in classrooms, visiting contractors, and on field trips, learning everything from astronomy to meteorology to flight mechanics to computers to spacecraft construction. If America was going to reach the Moon by President Kennedy's deadline, now just seven years away, the astronauts had to learn in gulps, not sips.

That applied to public relations, too. Meet-and-greets became commonplace, black tie functions the norm. Everyone in America, it seemed, wanted a piece of the astronauts. Once, Borman and Susan shared a limousine with a celebrity on their way to a gala sponsored by a wealthy Texas oilman.

"I'm Tony Randall," the man said.

"So nice to meet you," Borman said. "I really enjoyed your song 'I Left My Heart in San Francisco.'"

The actor did not appreciate being mistaken for the singer Tony Bennett. Borman did not appreciate the arrogance in Randall's indignation.

"To hell with him," Borman whispered to Susan.

As Borman settled in at NASA, it became clear to peers and management that he was a different breed, even among these unique men. He did not dabble in reflection, showed no patience for shades of gray. Mission came first, always, and if he sensed you were unqualified for a job or, worse, a bullshitter, he got your ass out of the way. He seemed unconcerned with NASA politics, blew smoke up no one's posterior, superiors included, and would not say, or do, anything he did not believe in. Some astronauts considered him arrogant or hard-headed, but all respected him, and few would have disagreed with Borman's own assessment—that he was among the best of the astronaut corps.

Like most astronauts, Borman was conservative politically. Yet he voted for Democrat Lyndon Johnson for president in 1964 because Borman believed strongly in racial justice and civil rights (it was a vote he'd later regret due to Johnson's policies in Vietnam). He was affected by Johnson's famous "Daisy" television commercial, aired during the campaign against Barry Goldwater, that juxtaposed a little girl against the mushroom cloud made by a nuclear bomb. The image disturbed Borman, yet he was ready, at a moment's notice, to drop the same kind of bomb on the Soviet Union if that's what America deemed necessary.

In 1964, Deke Slayton, the man in charge of crew assignments, teamed Borman with Jim Lovell to be primary crew for Gemini 7. The mission was planned as a fourteen-day Earth-orbital flight, the longest space mission ever attempted, intended primarily to test human endurance in space and to conduct a cascade of medical experiments.

During training, Borman and Lovell averaged more than twenty days a month away from home. When Borman got time off, he spent it with his family at home in Houston, taking Susan and his sons hunting and fishing. (Susan doubted she could bring herself to shoot a deer, but after Frank and the boys bought her a rifle, she had no trouble taking the shot. Frank never figured out whether she missed on purpose; to him, it meant everything that she tried.) To learn to water-ski, he and Susan checked out a book from the library, then took turns driving the boat, pages flapping in the wind. He loved how fast Susan took to it, even as he struggled. His boys delighted in how their father, a master of the skies, could barely swim. To make it to his sons' junior high football games, Borman pushed NASA's T-38 jets to their operational limits on Fridays after work, then ran to the hamburger stand Susan operated at the games, ready with his order in hand.

On Saturday, December 4, 1965, Susan and her two sons arrived at the VIP area at Cape Kennedy for the launch of Gemini 7. At 2:30 P.M., the Titan II rocket fired. As it rose in a column of white smoke and orange flame, Susan held on to her boys but looked away. Photographers captured the image—a good mother, a woman overwhelmed. Six minutes later, Gemini 7 was in orbit around Earth. Susan and her sons boarded a bus to the airport to go home. Out the window, Frederick and Edwin searched the sky for a glimpse of their dad's rocket ship.

Despite being confined to a cabin no larger than the front half of a Volkswagen Beetle, the longer Borman and Lovell flew, the more they liked each other. Every day, over and over, they sang "He'll Have to Go," a 1959 country ballad by Jim Reeves. "Put your sweet lips a little closer to the phone," they crooned; "Let's pretend that we're together all alone."

After eleven days in space, Borman and Lovell received visitors. Approaching like a white star, Gemini 6, which had just launched from Cape Kennedy, closed to within one foot of Gemini 7, proving that two ships could rendezvous in space (a necessary maneuver for flying a lunar landing mission, in which astronauts would use a lunar module to shuttle between an orbiting spacecraft and the Moon). Lovell burst out laughing when the Gemini 6 crew, Wally Schirra and Tom Stafford, flashed a sign to Borman: BEAT ARMY. Schirra, Stafford, and Lovell were Navy, and as a West Point man Borman had no choice but to take it.

By the time Borman and Lovell splashed down in the western Atlantic, they had set records for duration of flight (more than 330 hours, or 13.75 days), distance traveled (more than 5 million miles), and number of orbits (206). More important, they'd helped America take a major step toward the Moon by proving man could endure long stretches in space. The two weeks they'd spent was the maximum duration it was believed a lunar mission would require.

Borman was immediately made a full colonel, the youngest in the Air Force at age thirty-seven. A few weeks after splashdown, Susan wrote an article that was published in newspapers around the country. People had noticed how frightened she'd been during launch and the flight, and not everyone appreciated it—including some at NASA.

"These past weeks I had worn my feelings on my sleeve," she wrote. "Some said they were pleased to see an astronaut's wife willing to admit she was scared. Others, including some people in the space program, were critical because I failed to maintain the traditional stiff upper lip. 'For heaven's sake, wipe your tears. You're ruining my morning coffee,' one woman wrote. At one time, such criticism would have cut me deeply. But . . . I have come to realize you can't be all things to all people. So I decided not to pretend and not to try to hide my feelings—I decided to be myself."

Soon after Gemini 7's return, Borman received a telegram from West Point offering him a permanent professorship of mechanics. Susan loved the idea of returning to an idyllic life at West Point. But Borman said he couldn't do it—his heart was in flying, and he had a Cold War to help win. He would stay with NASA.

A year later, the tragic Apollo 1 fire occurred. Susan made it her mission to comfort and support her friend Pat White, the wife of one of the fallen astronauts. She visited the new widow every day, listening to her, holding her, and crying with her, trying to be strong as Pat kept repeating, "Who am I, Susan? Who am I? I've lost everything. It's all gone." At night, when Susan got home, she began to drink a bit, if only to quiet her nerves.

In the past, Susan had dealt with fatalities among Frank's colleagues the same way he did—by assuming it would never happen to him. But Ed White was different. He was a near-perfect physical specimen, even stronger than Frank, yet even he had been unable to get the spacecraft's hatch open during the fire. Frank told her that Charles Atlas himself couldn't have moved the hatch, but it was more than that to Susan. Ed White had been a West Point graduate, a devoted husband and father, and a committed patriot. He didn't screw around with muscle cars or other women. Which was to say he was just like Frank.

After eighteen months investigating the fire, testifying before Congress, and working on the Apollo command module redesign, Borman was offered the chance to be the commander of Apollo 8, man's first lunar mission. The flight was full of risks and unknowns, but it was where Borman had been pointing since he first soloed a single-engine airplane over the skies of Tucson. He hadn't known how that flight would end, either, but his instructor, Miss Bobbie, had believed he could go anywhere. Now, when he told Deke Slayton he would go to the Moon, he believed it, too.

Chapter Six

●

JUST FOUR MONTHS

ASTRONAUTS SCHEDULED TO FLY TO THE MOON IN JUST FOUR months should have been training in NASA's command module simulator, a ground-based model of the real thing. But, like most everything else connected to Apollo 8's new mission, it wasn't yet ready.

Borman, Lovell, and Anders settled on what their responsibilities would be for the mission, each according to his own experience and to his role on the flight. Borman would focus on the boosters and abort systems, the trajectory, and piloting the spacecraft. As commander, he would also be in charge during the flight, overseeing the crew and assuming responsibility for mission success.

Lovell would be the command module pilot, in charge of navigation. He would use the spacecraft's sextant, an optical instrument similar to those used on board sailing ships through the centuries, to measure angles between the Sun, Moon, and stars. (Primary navigation would be done by computers and Mission Control personnel, but Lovell needed to navigate, too, in case of technical failure on the ground or a complete loss of communications.) He would also map lunar landmarks and scout candidate areas for future landings. To learn the new guidance system, Lovell needed to spend time at MIT's Draper Lab, where he would practice sighting stars by focusing on the bright white light coming from atop a tall insurance building across the Charles River.

Anders would be the systems engineer, responsible for understanding how the highly complex spacecraft functioned. He had to master every switch, dial, lever, and gauge in the command module, where the astronauts would live for six days. He needed to have a thorough understanding of the service module attached to its base, which housed the systems for electric power and life support, and propellants essential to making the journey. There were thousands of intricate parts and connections and operations, and Anders had to make sure they all worked. He would also be in charge of photography, chronicling the flight on still and movie film. To this end, Anders fought to bring a 250 millimeter Zeiss Sonnar telephoto lens aboard. It was giant and heavy, but he had a feeling he'd need it.

Anders had come to change his thinking about Apollo 8's new mission in the weeks since it had been conceived. He'd been disappointed when told his crew would go to the Moon but wouldn't land there, given that it required him to give up his training as a lunar module pilot and become a command module specialist instead. On future missions, he'd probably be the guy who stayed behind in the orbiting spacecraft while his two crewmates walked on the Moon. For a man who dreamed of collecting rocks from the lunar surface, that packed a wallop.

But then he'd gotten to thinking: Flying on Apollo 8 meant that he, Lovell, and Borman would be the first human beings ever to leave Earth, and the first to arrive at the Moon. And the first to see its far side. That was like being another Christopher

Columbus, and what more could a curious man hope for than that?

The astronauts weren't the only ones under the gun. Director of Flight Operations Chris Kraft and others began constructing a detailed flight plan, one that accounted for every hour of the six-day journey; even a wasted minute would be unacceptable, given the risk and opportunity. Kraft also began his own study of the spacecraft and flight support systems; Kraft wanted to understand the ship better than the astronauts did, so if anything faltered, he'd already have been through the emergency and worked out every possible solution in his mind.

Nearly everyone involved in Apollo 8 had to coordinate with other departments, linking arms across NASA and industry to form a massive, cohesive whole. The agency and private industry needed to work together to prepare the command module, mate the spacecraft to the Saturn V rocket, and move it all to the Cape. Mission Control in Houston had to coordinate with the Cape to work out countdowns and launch windows, with the Marshall Space Flight Center in Huntsville to determine the rocket's maneuvers and trajectories, and with the contractors and universities that would help make complex calculations. It also had to make sure every part and every system was built to specification and on schedule. Computers and software had to be built and updated, electrical wiring diagrammed and tested, and the tracking stations around the world—which would relay voice and data between the flying spacecraft and Mission Control in Houston—brought up to speed. All of this, and so much else, had to be finished in just over one hundred days, all while NASA prepared for the launch of Apollo 7 in just one month. If that flight wasn't near-perfect, Apollo 8 wouldn't go.

On Friday evening, August 23, the astronauts went home for a rare weekend off. Many of their neighbors were like them—astronauts or NASA employees, conservative politically, with front lawns and haircuts that were military short. Boys still said "Yes, sir" when speaking to adults, girls still wore dresses. When the network newscasts came on the black-and-white televisions that night (color was still a luxury for many), few in these neighborhoods recognized the country looking back at them.

Thousands of antiwar protesters had descended on Chicago for the Democratic National Convention, which was scheduled to start in three days. Gathering in parks and on the streets downtown, these protesters, most of them under thirty years old, intended to make their demands for peace known to the Democrats, and to the world.

Rumors as to the protesters' intentions had circulated for weeks. Word had it that these long-haired young people planned to dump LSD into the water supply, stage nude-ins at Lake Michigan, turn over cars and toss Molotov cocktails, run off with delegates' daughters. A siege mentality took root in Chicago's elders. Except for one rally in Grant Park, Mayor Richard J. Daley refused to issue permits for the protesters to march, gather, or camp out in parks. To enforce Daley's peace, twelve thousand Chicago police officers, armed with military gear, stood at the ready, backed by six thousand members of the Illinois National Guard and six thousand regular Army troops.

To the astronauts, Chicago seemed a universe away. They lived military lives, rarely intersecting with the counterculture. Like most astronauts, Borman, Lovell, and Anders found the lifestyle and tactics of hippies and the antiwar movement unbecoming, even unpatriotic. But they didn't dismiss these young people. Each of them knew that the powder keg that looked ready to ignite in Chicago hadn't formed overnight; tensions had been building since the start of the year, one that was shaping up to be among the worst in the nation's history.

Already, ten thousand or more young Americans had been killed in Vietnam, and 1968 wasn't nearly over. Antiwar demonstrations had erupted around America, racial tension had led to riots, student protests had turned bloody. In a nine-week span, between early April and early June, two of the country's most inspirational figures—Martin Luther King, Jr., and Robert Kennedy—had been assassinated. Swaths of the population no longer trusted government or authority or institutions. Even music seemed more political—and angry—than before.

Now all of the year's turmoil seemed to be coming to a head in Chicago. As the crew of Apollo 8 prepared to resume training after a weekend at home, two thousand demonstrators massed in Chicago's Lincoln Park. Many had nowhere else to sleep, but the city's curfew required them to disperse at 11 P.M. When the hour struck, police outfitted in gas masks and helmets moved in, firing tear gas canisters into the remaining crowd, clubbing and kicking whomever they could reach.

The convention opened the next day. Protesters marched on police headquarters, then redirected to Grant Park. At the convention, Daley promised, "As long as I'm mayor of this town, there will be law and order in Chicago." On television, 89 million Americans tuned in to see the direction the country might take.

On August 28, the Democratic Party voted against adopting an antiwar plank to its platform. The peace candidate, Eugene McCarthy, refused on principle to address the convention, and Vice President Hubert Humphrey secured the nomination. A crowd of ten thousand rallied in Grant Park. Tempers flared, and soon billy clubs and boots were flying. Rennie Davis, one of the organizers of the demonstrations, was beaten unconscious. Thousands began to march to the site of the convention, but they were turned back by National Guardsmen, some brandishing automatic weapons and grenade launchers.

That left the protesters outside the Conrad Hilton Hotel on Michigan Avenue, where they remained into the night. By the thousands, they shouted epithets and profanities at police, delegates, politicians—anyone in charge—and to many it no longer sounded like free speech or the expression of opinion, it sounded like America had burst, and the bile that discharged flowed uphill on one of America's most exclusive streets, seeking higher and higher levels so that Humphrey would hear it in his twenty-fifth-floor room, and LBJ would hear it at his ranch in Texas.

The police stood there taking the worst of it. In ordinary times, a person cursed at a Chicago cop at their peril. Yet peril seemed to be what the demonstrators wanted most. After thirty minutes, the police obliged them, smashing and clubbing and kicking and dragging anyone they could reach—demonstrators, onlookers, journalists—and it didn't matter that the network television cameras were filming or that people were yelling "The whole world is watching!" or that those in the streets weren't Vietcong or Soviets but the sons and daughters of fellow citizens; all that mattered for the next eighteen minutes of brutality and mayhem was that something had fractured in America and no one had any idea how to stop it, and after order was restored there still seemed to be cries coming from the streets, even though there was no one left to make them. Among the millions who watched the unedited footage on television, there hardly seemed a soul among them—rich or poor, young or old, left or right—who didn't wonder if America could be put back together again.

On Sunday, September 8, the crew of Apollo 8 flew their T-38 jets from Houston to Cape Kennedy in Florida, checked in to the Holiday Inn at Cocoa Beach, and prepared to die.

Often.

In the morning, they would start training in the command module simulator, an Earth-based model of the Apollo spacecraft they would pilot in December. Housed in nondescript buildings at the Manned Spacecraft Center in Houston and at Cape Kennedy in Florida, the machines were highly accurate mock-ups of the real thing, their cabins outfitted with every switch, lever, dial, gauge, light, alarm, circuit breaker, and read-out the astronauts would use on the lunar journey. Everything worked—the simulator itself didn't move, but optics could be projected onto screens, navigational information displayed, sounds played over speakers, and lights flashed. The seven hundred–plus manual controls functioned just as they would during an actual mission. An astronaut could spend years poring over diagrams and schematics, but he would never know a complex machine like the command module without spending time inside its landlocked twin.

Any segment of a mission could be replicated, any situation reproduced, any scenario played out. The astronauts could "fly" the simulator just as they would the real spacecraft, working through segments of their lunar journey in real time. Any mistakes and they'd know it. And they wouldn't be the only ones.

Seated outside the simulator were a set of NASA employees who served as instructors. They were the ones who programmed scenarios into the simulator, and who watched the astronauts' every move on their consoles, ready to make critiques and corrections.

Leading the team of instructors was the Simulation Supervisor, or SimSup. One of his jobs was to teach the astronauts the correct sequences and procedures for every part of the flight, from liftoff to lunar orbit to splashdown. His other job was to kill them.

Space flight was inherently complex and unpredictable—crews were nearly certain to encounter problems with the rocket and spacecraft during their mission. To give them a fighting chance, the SimSup would unleash an arsenal of emergencies, failures, malfunctions, and conflicts into the simulation, forcing the crew to learn to survive, showing them the consequences of every wrong move. It would do no one any good to take it easy on them. Only by theoretically endangering the lives of the men inside the simulator could the SimSup hope to save them during actual flight. In this way, the best SimSups had a streak of the devil inside them.

Flight controllers and others involved in the mission would also work with the SimSup and instructors. Astronauts in the simulator at the Cape would be able to talk to controllers in Houston, to launch specialists in Florida, even to ground stations in Australia. The simulation was nearly as elaborate as the actual flight. The crew of Apollo 8 knew they had to be ready.

Borman, Lovell, and Anders ate a predawn breakfast together, joking about how close they'd have to sit in the simulator, threatening whoever dared show up without brushing his teeth. The Sun was just rising when they arrived at the command module simulator at Cape Kennedy. It was Monday, September 9, 1968, less than fifteen weeks before Apollo 8's scheduled mission. Despite the early hour, the room was crowded with flight controllers and technicians, many of whom were in their twenties, some just out of college. At age forty, Borman and Lovell were nearly twice as old as some of the men gathered around the simulator; even Anders, at thirty-four, seemed an elder statesman here.

As test and fighter pilots, the astronauts had flown cutting-edge machines, but even they needed time to process the sight of the Apollo simulator. Standing about twenty feet high, it was a hodgepodge of sharp-cornered modules that appeared jammed together by cubist painters, jazz musicians, and mad scientists. There seemed no front or back, or even up or down, just shapes. Hundreds of cables dangled from the contraption like dreadlocks, while two narrow staircases—one circular, the other straight—led inside, or at least somewhere. Bracketing the structure were consoles of computers, instruments, and monitors for the instructors. Fluorescent white light bathed the room.

After a briefing, a technician directed the crew to the straight staircase, a steep incline of fourteen carpeted steps with spaghetti-thin handrails that led to the simulator's hatch. For the most part, the astronauts would not need to wear their flight suits in the simulator, which was good news on this day, since their flight suits still hadn't been made.

Once inside the cabin, the crew lay back in their seats (also called couches, since they supported the men's bodies from head to toe), Anders on the right, Lovell in the center, Borman on the left. Anders stared at the panels of lights and indicators that were flickering to life, knowing it would take the entirety of his focus over the next hundred days to learn to ride the real thing out of this world.

Borman looked at Lovell and Anders. He'd always been a sharp student of character, and as he sat there, he believed he had the best crew ever assembled by NASA.

The closing of the hatch echoed inside the cabin.

"All right," Borman said to his crewmates. "Let's learn how to go to the Moon."

A week later, on September 14, the Soviet Union launched an unmanned spacecraft toward the Moon. Both the Americans and Soviets had sent probes to the Moon in the past, but this one, called Zond 5, was different, because the Soviets intended to get it back.

No spacecraft had ever come near the Moon and returned safely to Earth. If the Soviets could pull it off, it would represent a major leap forward, and a clear signal they intended to send men to the Moon in early December, their best launch window, and two weeks before Apollo 8's scheduled lift-off.

Streaking out of Earth's atmosphere, Zond 5 carried tortoises, wine flies, mealworms, and other living organisms. Strapped into the pilot's seat was a five-foot-seven, 154-pound mannequin, its sensors absorbing radiation data. With modifications, the same ship could carry two cosmonauts.

The day after Zond launched, NASA chief James Webb announced his resignation, effective October 7. (Earlier in the year, upon learning that President Johnson wouldn't seek reelection, Webb had decided to step down.) Until then, Thomas Paine would continue as deputy administrator, then assume the reins in Webb's place. Word of Webb's resignation surprised the NASA brass. Most considered him a giant, as responsible as any person, Kennedy included, for making the American space program world class. But there was a silver lining. Webb had never been fully on board with the plan to send Apollo 8 to the Moon in December. Paine always had been.

A day later, good news arrived for Apollo 8. The Saturn V rocket had passed its design certification review, meaning the fixes and modifications made by von Braun and his engineers after the booster's troubled test flight on Apollo 6 in April had been judged to be effective. Even the violent pogo problem seemed to have been tamed. Pending a few final checkouts, the rocket looked ready to launch three astronauts to the Moon.

The crew of Apollo 8 spent much of the next day, September 18, in the command module simulator in Florida. They were joined, as they often would be during training, by their backup crew, Neil Armstrong, Edwin "Buzz" Aldrin, and Fred Haise. By Slayton's assignment scheme, backup crews became primary crews three flights later. That meant Armstrong, Aldrin, and Haise would be prime crew for Apollo 11.

Late that night, while the astronauts slept, the famed British astronomer Sir Bernard Lovell (no relation to Jim Lovell) reported that a massive radio telescope in England had tracked a Soviet spacecraft (Zond 5) as it passed within a thousand miles of the Moon. Further, it appeared that the ship was now making a return journey to Earth. Lovell concluded the Soviets intended to recover the craft. "Once they have achieved this," he said, "we can anticipate that they will put a man in one."

The next night, observers picked up a different kind of signal being broadcast from Zond 5. This time, they heard a Russian voice.

No one believed a cosmonaut to be aboard Zond 5, but the man calling out the ship's instrument readings was as real as the spacecraft itself. His voice, and others heard later, belonged to cosmonauts and were being transmitted live from the Soviet Union to Zond 5, then beamed back to Earth by the spacecraft, all by way of practice for the real thing. Soviet intentions were clear. A manned lunar mission was coming very soon.

But Zond 5 wasn't home yet.

On September 21, the spacecraft collided with Earth's atmosphere at a speed of 24,600 miles per hour. In seconds, deceleration forces reached between 12 and 18 g's, a punishing (but survivable) load for properly trained humans (1 g is equal to the force of gravity at Earth's surface, 2 g's is equal to twice the force of gravity at Earth's surface, and so on). For three minutes, Zond 5 raked through increasing resistance until it plummeted through a darkened sky toward the Indian Ocean. At an altitude of about 20,000 feet, its single parachute deployed, leaving the craft, still glowing from the heat generated by reentry, in a final ride to the water. Still alive inside the capsule were the tortoises, just 10 percent lighter for their near-week in space. Several fly eggs had hatched. It had been a rough return, but the bottom line was unmistakable: living creatures had survived a round-trip to the Moon.

In England, Sir Bernard Lovell told reporters that Russia had regained the lead in the race to send men to the Moon and that Zond 5 "makes it highly probable that a Russian will get a close-up look at the Moon quite a long time before an American does."

At NASA, the preparations for Apollo 8 grew even more intense. While the astronauts trained for twelve or more hours a day, the spacecraft was moved to the Vehicle Assembly Building at Cape Kennedy, where it was mated to the Saturn V rocket.

Voices of opposition to the Apollo 8 mission began to ring out. In a September 24 editorial, *The Washington Post* warned,

"Our program . . . ought to move at its own pace. If that pace is sufficiently rapid to bring American astronauts to the Moon first, fine. If it is not, so be it. The Russians will deserve the honor and praise they will win if their men make the first landing. In space exploration, it is more important to do things right than to do them first." In a letter to Webb, astronaut Buzz Aldrin's father, himself a former Air Force colonel and an aviation pioneer, wrote, "I do not favor a manned flight of Saturn V until the changes being made have been proven. What is the value of risking lives at this stage? You really need less yes-men in the space program."

On October 9, a sky-high bay opened at the Vehicle Assembly Building, revealing the gargantuan Saturn V, white with black patches and streamlined to a narrow point at the top, an elegant monster fifteen stories taller and five times heavier than the Saturn IB the crew of Apollo 7 were scheduled to ride into Earth orbit. Only a rocket with that kind of size and power could lift a payload as heavy as an Apollo spacecraft bound for the Moon (although most of the rocket would fall away in the first few minutes of flight, and the rest of it a few hours later).

For several minutes, the Saturn V stood and gleamed in the Florida sun. And then it started to move.

The rocket stood atop NASA's Crawler-Transporter, the world's most powerful tractor. Powered by two sixteen-cylinder engines with a combined 5,500 horsepower and sixteen locomotive traction motors, the Crawler-Transporter was the largest self-powered land vehicle in the world. By itself, it weighed 6 million pounds, and it could move payloads in excess of 12 million pounds. The 131-foot-long vehicle rode on eight tank-like tracks—two on each corner, each pair the size of a Greyhound bus—and could deliver its payload to within inches of its target destination. Its top deck was the size of a major league baseball infield. Carrying the Saturn V to the launchpad, it would cruise at one mile per hour.

Engines grinding, the tractor moved the Saturn V and its tower out of the white Vehicle Assembly Building and into daylight. A slender 34-foot-tall launch escape system sat atop the rocket and seemed to scrape the nearly full Moon hanging in the sky. Soon the structure was moving down the road toward Pad 39A, a journey of about three and a half miles, where the spacecraft would undergo exhaustive testing, verifications, and countdown rehearsals until the December launch. A man in white shirtsleeves and a black tie operated the Crawler-Transporter from inside a control cab, while several engineers wearing hardhats rode atop various platforms on the tractor. Like the red fire engine that drove beside them in case of emergency, these men appeared to be toys in the shadow of these machines.

As the Apollo 8 hardware made its reptilian crawl, it might have been easy to forget that in just two days, Apollo 7 would launch on an eleven-day Earth orbit mission designed to test the Apollo spacecraft and systems. That flight, historic in its own right, and NASA's first since the fatal fire, had to be near-perfect for Apollo 8 to get its green light for December.

Apollo 7 sat atop its Saturn IB rocket on October 11, 1968, a 20-knot easterly wind blowing against the spacecraft and into commander Wally Schirra's instinct. Mission rules prohibited launching into winds that could push a spacecraft back onshore during an abort—the ground could be a deadly hard landing spot compared to the ocean—but that was just the kind of wind whistling at Cape Kennedy during the countdown. Lying on his back alongside crewmates Donn Eisele and Walt Cunningham, Schirra grew furious that NASA seemed determined to fly despite the hazard he perceived. He argued against launching until about an hour before lift-off, when he realized it was too late to call things off.

Just after 11 A.M., with the wind still howling, Apollo 7 launched successfully. All went well until the second day, when

Schirra came down with a head cold—a condition made even more uncomfortable given that noses don't run in zero gravity. To make matters worse, the crew struggled with their biomedical equipment, strained to see out their windows, and were forced to pump waste water manually from the spacecraft. When Mission Control asked about the live television broadcast scheduled for that day, Schirra made it clear where he stood.

"You've added two burns to this flight schedule, and you've added a urine water dump, and we have a new vehicle up here, and I can tell you [at] this point TV will be delayed without any further discussion until after the rendezvous."

It got worse. Though the spacecraft was functioning beautifully, the astronauts' attitudes were breaking down. On day seven, Cunningham said to controllers, "I'd just like to go on record here as saying that people that dream up procedures like this after you lift off have somehow or another been dropping the ball for the last three years. . . . It looks kind of Mickey Mouse." On day eight, Schirra said, "I wish you would find out the idiot's name who thought up this test. . . . I want to talk to him personally when I get back down."

In Houston, Kraft and Slayton were seething. Not only were Schirra and his crew nearly insubordinate, they were doing it for the public to hear. At a press briefing, a reporter said, "I've covered sixteen flights, and I don't recall ever finding a bunch of people up there growling the way these guys are. Now, you're either doing a bad job down here, or they're a bunch of malcontents. Which is it?"

Apollo 7 splashed down eleven days after lift-off. Every mission objective had been achieved, and more. The spacecraft had worked beautifully. The SPS engine, so critical to a lunar journey, had performed well. By virtually every measure, the flight had been nearly perfect, and it would open the door to Apollo 8's flight to the Moon.

Many attributed the negative behavior by the crew of Apollo 7 to the constant discomfort from their head colds. Others wondered if Schirra had been terrified by the Apollo 1 fire. The commander of that mission, Gus Grissom, had been Schirra's next-door neighbor. Schirra had been Grissom's backup pilot for the flight. Long after the fire, Schirra had told people, "We all spent a year wearing black arm bands for three very good men. I'll be damned if anybody's going to spend the next year wearing one for me."

Despite the technical brilliance of the mission, Kraft wouldn't abide insubordination, even if it was born of legitimate fear; he determined that none of Apollo 7's crew would ever fly again for NASA. He felt differently about the crew of Apollo 8. Borman, Lovell, and Anders were consummate professionals, as rock steady as they came. He was certainly grateful for Lovell. Kraft had been ringside for Gemini 7, the grueling fourteen-day mission during which Lovell remained unflappable, even during problems that might have threatened the flight's survival. Equally important, Lovell was as likable and optimistic a fellow as there was in the astronaut corps, and on man's first journey away from his world, there could never be too much of that.

Chapter Seven

●

JIM LOVELL

EVERY SATURDAY WHEN HE WAS FIVE YEARS OLD, JAMES ARTHUR Lovell, Jr., went to the movies, always to see a Western. Sometimes he went with his father, but the best times were when he went alone. Walking untethered through Philadelphia as a little boy, he was free to discover new neighborhoods, invent new routes, pass strange faces, navigate a giant world by himself.

Born in 1928, Jim grew up in the teeth of the Depression, but his father had work and the family didn't want for much. All of that changed around the time Jim reached fifth grade; his parents separated, and not long after, his father died in an automobile accident. Needing to support herself and her young son, Blanch Lovell moved to be near her brother in Milwaukee, Wisconsin, and took a job as a secretary for modest wages. By 1940, she and Jim were living in a tiny one-room apartment, their kitchen jammed into a closet, using a single toilet shared by everyone who lived on the floor.

In 1940, an American kid could hardly walk into a drugstore without seeing a new kind of flying machine streaking across magazine covers: the rocket. Jim couldn't get enough of the fins and flames and faraway planets painted in full color by the magazine's visionary artists, or the stories of what these machines could do. Rockets didn't just take a person from point to point, like airplanes. They flew into the future.

Jim wanted to fly there, too. Soon he was reading books by the founding father of rocket engineering, Robert Goddard. The idea that these machines could reach beyond Earth's atmosphere lit up Jim's dreams. He read Jules Verne's novel *From the Earth to the Moon,* and its sequel, *Around the Moon,* which tell of three adventurers who build a nine-hundred-foot space cannon that launches them in a projectile around the Moon. During their journey, the men avert a deadly asteroid strike, jettison a dead dog out the window, and succumb to a mysterious force that causes them to dance and sing. They also glimpse the far side of the Moon, a view unavailable from Earth. Now, seventy-five years after Verne had penned his science fiction masterpieces, rocket engineers were saying that an actual trip to the Moon might be possible. Jim paid attention to that.

By the time he began at Milwaukee's Juneau High School, Jim had determined to learn all there was to know about rocketry. He discovered a report written by Goddard and published in 1919, *A Method of Reaching Extreme Altitudes,* and was fascinated by the vision in Goddard's mathematical calculations and his thinking about rocket fuels. *The New York Times* had ridiculed Goddard for suggesting that a rocket could operate in the vacuum of space or carry payloads to the Moon. "He only seems to lack the knowledge ladled out daily in high schools," the newspaper wrote. Goddard responded by saying, "Every vision is a joke until the first man accomplishes it; once realized, it becomes commonplace." To fourteen-year-old Jim Lovell, Goddard had more than vision. He had courage.

In June 1944, the summer after his sophomore year, Jim took

a job baling hay on a farm in Plymouth, Wisconsin, an hour north of Milwaukee. The pay wasn't much, about ten cents an hour, but after work, he and the other young farmhands piled into trucks for a ride to the local lake, where they swam late into the night. Lying on his back beneath crystalline Wisconsin skies, Jim could pick out the Big Dipper, the North Star, and Cassiopeia, all of which he'd learned to use for navigation. All through the summer, celestial bodies moved across Jim's nights, calling to him from their black canvas.

Working as a server of hot foods in the cafeteria during his junior year, Jim spotted Marilyn Gerlach, a pretty freshman he'd admired all year. He decided to ask her to the prom.

"I don't know how to dance," she said.

"I don't either," Jim replied. "We'll learn together."

He went to Marilyn's house, introduced himself to her parents, and played his record albums in her living room as they practiced their dance moves. Prom night came, the dance floor shook, and Jim and Marilyn became an item.

Near the end of his junior year, Jim and some friends planned to build a rocket. Gathering cardboard mailing tubes for the body and #10 tin cans for fuel tanks was a cinch; finding rocket fuel was another matter. Jim got a formula from his chemistry teacher, then found a company in Chicago that sold the ingredients to make the fuel, but when he arrived something seemed amiss—the place, located in a tall building downtown, looked more like an attorney's office than a hardware store. When Jim placed his order, the receptionist arched an eyebrow.

"You want sulfur, potassium nitrate, and charcoal?"

"Yes, ma'am, just a few pounds."

She asked for Jim's name—his full name—then summoned a man from the back.

"Do you know what those chemicals make when mixed together?" the man asked.

"Yes, sir. Rocket fuel."

"No, son. That's gunpowder."

No one looked more surprised than Jim. But he told the man he was still willing to buy it.

The man, however, was not willing to sell. For one, he told Jim, the company sold its chemicals by the truckful. Second, Jim was seventeen. Third, fourth, fifth, and sixth, Jim was seventeen.

Back in Milwaukee, Jim's teacher got a kick out of the story, then helped him and his friends find the chemicals in appropriate quantities. A few days later, a three-foot rocket took shape, complete with wooden nose cone and fins, and a fuse made from a soda straw. Protected by a welder's mask, Jim took the creation to an open field, lit the fuse, and ran with pals for cover behind rocks. Across the way, Marilyn watched from a safe distance. On ignition, the rocket screamed into the sky, leaving a trail of crooked smoke as it climbed eighty feet before exploding and raining down blackened shards of cardboard tubing. Somewhere, Robert Goddard was smiling.

Toward the end of Jim's senior year of high school in 1946, he visited a Milwaukee fairground. There, he witnessed a wonder. On display, close enough to touch, was the spent engine of a captured V-2 German rocket, the one designed by Wernher von Braun and used by the Nazis to attack European cities at the end of World War II. The rocket could travel 200 miles, carry a ton of explosives, reach an altitude of 50 miles, and attain speeds of more than 3,300 miles per hour.

Electrified by the encounter, Jim wrote to the American Rocket Society, which had already existed for twenty years, and asked for advice on careers. They replied with a friendly letter telling him that universities didn't yet offer majors in rocket technology, but that he'd be well served to take college courses in thermodynamics, aerodynamics, and mathematics. To that end, the society advised, the Massachusetts Institute of Technology or the California Institute of Technology would make excellent choices.

Jim had no money for college. His mother encouraged him to apply to the Naval Academy at Annapolis, from which his uncle had graduated in 1913, and Jim did that, but the best the Navy could offer was a position as a third alternate on the admissions waitlist. The Navy, however, did need pilots. To get them, they were willing to pay for a student to go to college. If that student did well, the Navy would make him a military aviator. All of it would be paid for by Uncle Sam.

The idea of being in command of his own aircraft thrilled Jim. He signed up for the Navy program and enrolled at the University of Wisconsin–Madison, where he loaded up on mechanical engineering courses and saved most of his fifty-dollar-a-month stipend for weekends when Marilyn came to visit.

After two years in Madison, Lovell moved to Pensacola for flight training. Halfway through the preflight segment, he received orders to report to the Naval Academy; a few days later, Lovell was just another plebe at Annapolis. None of his credits from Wisconsin was eligible for transfer. He was starting from scratch.

In November, Lovell invited Marilyn to the Army-Navy football game in Philadelphia. By that time, he knew he wanted to marry Marilyn, and he asked her to move out east near the Naval Academy. At the time, Marilyn was attending a teacher's college at home in Milwaukee, but she pulled up stakes and moved to Washington, D.C., where she enrolled at George Washington University and took a job at Garfinckel's Department Store. And started dating a medical student.

Lovell could hardly blame her. He was so busy at Annapolis he hardly had a chance to call, let alone take out, his girl, and even when midshipmen got liberty for a night on the town, there were curfews and other style-cramping rules. Still, the med student didn't last long; he didn't like it when Marilyn wore Lovell's class crest on her sweater one night.

Things got easier for Lovell and Marilyn in his third year, when the Academy allowed him more liberty. After dinner one night, he walked her to a jewelry store, where they admired a selection of engagement rings.

"Do you like that one?" Lovell asked.

"Do you want me to have one of those?" she replied.

"After seven years, I don't want anything else," he said.

Lovell tore into his final year at Annapolis. Only fifty in the class of nearly eight hundred would be assigned to flight school right away, and finishing strong depended in part on the quality of one's senior thesis. Most played it safe, writing on naval history or tactics, but Lovell took a leap into the theoretical. Working long into the nights, and with Marilyn as his typist, he put together a study of the development of the liquid-fuel rocket engine, a paper that didn't just analyze the state of the art but made predictions that sounded more Jules Verne than midshipman.

"The big day for rockets is still coming," he wrote, "the day when science will have advanced to the stage when flight into space is reality and not a dream. That will be the day when the advantage of rocket power—simplicity, high thrust, and the ability to operate in a vacuum—will be used to best advantage."

Even in 1952, talking about combustion in a vacuum could seem ridiculous to the uninformed. But Lovell's vision never wavered. When the paper came back it was marked A minus. Lovell graduated at the top of the class on June 6, 1952. Later that day, he and Marilyn were married at St. Anne's Episcopal Church in Annapolis.

As he'd long hoped, Lovell was chosen to attend the Navy's flight training program. He returned to Pensacola, this time as an officer (ensign), not a midshipman. A year after they arrived, Marilyn gave birth to the couple's first child, Barbara. Two months later, in February 1954, Lovell earned his wings and was ordered to the Naval Air Station at Moffett Field in California, a few miles from Palo Alto.

He was assigned to VC-3, a squadron that supplied fighter pilots trained in night operations to aircraft carriers in the Pacific. Few assignments tightened a flier's throat like landing a jet on a darkened deck just a few hundred yards long in the dead of night—a tiny moving runway on roiling seas. Small errors could become deadly mistakes.

One moonless night in early 1955, after launching from the deck of the USS *Shangri-La* in an F2H Banshee jet fighter off the coast of Japan, Lovell embarked on his first combat exercise over foreign waters. Bad weather had prevented takeoff for the last of the four fliers in Lovell's patrol. The jets that had already launched were ordered to circle the ship until they burned down their fuel, then land. Cloud cover forced the Banshees to stay just 1,500 feet above the choppy seas.

Lovell banked to join his teammates in formation, but when he reached the rendezvous point he was alone. His automatic direction finder (ADF) indicated he was heading straight for the carrier, but the ship wasn't there. The other pilots reported that they were already circling the *Shangri-La*. Something in Lovell's navigation had gone wrong.

He checked his ADF and confirmed he was locked on to the ship's frequency. But he was not, in fact, locked on to the ship. His instrument had instead picked up a Japanese tracking station broadcasting on the same frequency. Without knowing it, he was following that signal, in total darkness, to the coast.

Sensing that something was wrong, Lovell banked 180 degrees to look for his wingmen. All he found were empty skies. He reached for the flight plan to make sure his radio numbers were correct, but it was too dark to read the small print by the jet's ambient light. Lovell had a solution for that. He'd designed and built a small light, which he'd carried along and now plugged in.

Circuits blew. Every light in the cockpit died. The airplane turned as dark as the night.

Lovell had to make a choice. He could ask the carrier to turn on its lights, an embarrassment from which he might never recover. Or he could continue following the signal, hoping to find the ship—or Japan—before he ran out of fuel.

In the end, he chose neither. And it was all because of green.

Lovell saw the color barely glowing in the water below him. He knew that algae could be made luminous when churned by the spinning propellers of a powerful ship, so he decided to follow the faint flare in the water. Several minutes later, he located his wingmen, who set down, first one and then the other, on the deck of the *Shangri-La*. Next it was Lovell's turn to land, and even though he'd found home, his cockpit was still without lights. He couldn't tell his airspeed and altitude without being able to read his instruments.

But he still had a penlight, and he flipped it on, then put it in his mouth to cast its tiny beam on the instrument panel before him. Believing himself to be about 250 feet above the water, he descended toward the carrier, only to see his wingtip's red light reflect on the water no more than twenty feet below, a split second from impact. Lovell cranked back on the stick and jammed forward the throttle, sending the Banshee howling skyward and just clear of the side of the *Shangri-La*'s deck.

Heart pounding and mouth dry, Lovell now had to turn back and try again. This time, he came in high but, despite frantic don't-do-it signals from the landing officer, figured he'd never get a better chance to make the flight deck, given his limited ability to read his instruments. Plummeting downward, he thudded onto the deck and skidded forward, one of his tires blowing before the carrier's last arresting wire grabbed the jet's tailhook and jerked the plane to a halt.

Lovell's legs shook so badly he could barely climb from the cockpit. But that experience only confirmed how he felt about death. To him, the only thing guaranteed to a person was the moment. It was the only time one knew he would be there to take in the trees and the sun and the stars, to meet people, make friends, fall in love. But a person couldn't be in the moment if he worried too much about the future. That meant in order to live, he couldn't worry about dying. The day after Lovell's wild flight, he climbed back into the cockpit and took off again. This time, he put the airplane back down just where it belonged.

While Lovell was training, Marilyn gave birth to their second child, James Jr. A few weeks later, Lovell watched transfixed as America's lead rocket designer, Wernher von Braun, appeared on a nationally televised Disney special, *Man in Space*. Von Braun showed viewers a prototype space suit like the one Americans would wear "when we make the trip to the Moon," and he revealed his model for a four-stage orbital rocket ship—about the coolest thing any of the 42 million people watching the program had ever seen.

For the next two years, Lovell flew jets at sea and trained pilots while Marilyn raised the children in California. In 1957, he applied to the test pilot school at the Naval Air Test Center at Patuxent River, Maryland. The job of testing experimental aircraft built with the most advanced technology seemed a natural fit. Marilyn backed his decision and packed the Lovells to go.

The training program at Pax River lasted for six months. At graduation, Lovell ranked first in his class. His gift from Marilyn was a new daughter, Susan, making them a family of five. Soon after, in 1958, Lovell and some other pilots at Pax River received a telex from a new government agency, the National Aeronautics and Space Administration, ordering them to a meeting in Washington, D.C. They were to dress in civilian clothes and not tell anyone, including family, where they were going, or even that they were going at all.

When he arrived, Lovell joined dozens of other military pilots for a briefing in a government office. Robert Gilruth, head of NASA's Space Task Group, explained that the agency was looking for astronauts for Project Mercury, a program designed to put a manned spacecraft into orbit around Earth and recover it safely. He laid out NASA's vision, talking of rockets and capsules and head-spinning speeds. Think things over tonight, Gilruth told the men, then report back tomorrow for more.

Some participants questioned the wisdom of abandoning a Navy career to enroll in an astronaut program that hadn't yet started and might not even exist in a few years. As for Lovell, he could hardly believe his luck.

Several days later he was in New Mexico, enduring six days of torturous physical exams. At the end, doctors failed him—or rather, his body—for having a bit too much bilirubin, a pigment produced by the liver and found in bile. They didn't think the level dangerous, but that didn't matter; what they seemed to demand was physical perfection. "You're finished," they told Lovell, and no matter how forcefully he explained their mistake, the doctors wouldn't reconsider. "I could spell 'rocket' before these guys ever heard the term," Lovell muttered as he walked away. Back at Pax River, Marilyn couldn't remember having seen her husband so discouraged.

A short time later, Lovell received orders to report to the next phase of astronaut testing. He knew he'd been rejected, and that the orders had been issued by mistake, but he seized his chance to get back in the game, even if by clerical error. He flew to Wright-Patterson Air Force Base in Ohio and took the last bed in quarters. The next morning, just as the miracle seemed complete, an Air Force test pilot named Gus Grissom showed up and apologized for being late. Lovell was again heading back home with nothing to show for his dreams but a little extra pigment in his liver.

For the next three years, Lovell continued testing aircraft and teaching students at Pax River. It was there that the nickname Shaky was bestowed upon him, not just because no decent pilot would want such a moniker, but because the easygoing Lovell was among the least shaky men around.

By 1962, Project Mercury was nearing its end and NASA needed new astronauts. That summer, the Navy asked if Lovell would like to apply. No one seemed to remember that he'd been medically disqualified, and Lovell could find no good reason to remind them.

Again, Lovell went through the testing. As a teenager, he'd seen the engine of a Nazi V-2 rocket designed by Wernher von Braun. As a young pilot, he'd watched von Braun tell the nation how America would go to the Moon. After what seemed like forever, Deke Slayton called and asked if Lovell would like to ride the great engineer's newest rockets for himself, and Lovell's answer could be heard all the way to Milwaukee. He was officially one of NASA's New Nine.

Lovell was introduced to NASA's eight other new astronauts at the Rice Hotel in Houston. After dinner, he gave his first comment as a spaceman, telling his hometown newspaper, the *Milwaukee Sentinel,* that America would be first to the Moon, "and I want to be on the first team."

In Houston, Lovell took up residence in old World War II barracks at Ellington Air Force Base, where residents lived four to a unit and had bedsheets for walls. His family soon followed and before long, Marilyn found a house to rent in a nearby suburb. For seven-year-old Jay Lovell, that was the perfect setup: Ellington was just a few minutes away, and his dad was only too happy to take him along to the airfield to watch the training he and the other astronauts were doing in their T-38 jets. Jay loved it when his dad retracted the landing gear and kept flying just a few feet off the ground, but he stood awestruck when his father once did something radically different. On that day, Lovell pointed the jet straight up after takeoff, and as Jay watched asphalt fly and ground crew scurry, he could see that his dad was aiming right for the Moon.

As Lovell learned his way around his new job and his new city, Marilyn settled the family into their new home in a small Houston subdivision called Timber Cove. The sudden celebrity that came with being an astronaut startled both of them. People even recognized Marilyn around town. Lovell understood the slight resentment he and some of the other new astronauts detected from the Original Seven; the new guys hadn't even entered a spacecraft yet, so who were they to soak up America's adoration?

Soon enough, though, the veterans warmed to the rookies. Once, when Lovell needed a ride, Alan Shepard told him to jump into his brand-new 1963 Corvette, a car that had come complete with the astronaut's name engraved on a plaque. Shepard had the top down and opened the throttle on I-45 in Houston, showing Lovell what speed really meant. "Boy, how much do these things cost?" Lovell asked. "If you gotta ask, you can't afford one," Shepard replied. Lovell made a mental note: *Get one.*

In 1964, Lovell got his first assignment, as one of the two-man backup crew for Gemini 4. His partner would be Frank Borman, whom he'd met during medical exams of astronaut hopefuls. Slayton had named Borman the commander, Lovell the copilot. To Lovell, that didn't seem quite fair; they were about the same rank, and he couldn't see why Borman was any more qualified to assume responsibility for a flight than he. But no matter who was commander, there was wonderful news in the assignment. By Slayton's scheme, Lovell and Borman would be the primary crew for Gemini 7, a two-week Earth-orbital flight, the longest mission ever planned by the space agency. In a matter of months, James Arthur Lovell, Jr., would be going into space.

To some, the pairing of Borman and Lovell might have seemed curious—even doomed. Borman didn't bother with space dreams, spent no energy imagining the heavens. He'd come to NASA for a single purpose—to help America defeat the Soviet Union. In meetings or in training, he could come off as brash or bullheaded if he believed you to be impeding the mission; sometimes he'd walk out on a discussion, even over drinks after work, if he sensed bullshit in the air. That kind of directness earned Borman almost universal respect, but not everyone liked him for it.

Lovell seemed Borman's opposite. He had ridden a dream—of

exploring the cosmos and flying rockets to new worlds—from childhood all the way to NASA. And while it would please Lovell to beat the Russians, he mostly thrilled to the idea of going places forever thought unreachable and reporting back to the world about what he'd seen. Few could deny Lovell's abilities as a thinker or a pilot, but it was his warmth and friendliness that people remembered most.

And yet, from the day they began working together, Borman and Lovell seemed a natural match. Each respected the other's abilities, work ethic, intellect, and piloting skills. And they made each other laugh. To many, it seemed the men had been friends since boyhood.

In early June 1965, Lovell packed Marilyn and their three kids into the car and drove from Houston to the Cape to watch one of the Gemini launches in person. Asleep in bed one night at the Cocoa Beach Holiday Inn, Lovell was awakened by the sound of his wife munching saltines.

"What's going on?" he asked.

"I hate to tell you this," Marilyn said, "but I think, I mean I know . . . I'm pregnant."

It was great news but could not have been more awkwardly timed. Marilyn was due around the time Lovell was scheduled to fly on Gemini 7 in early December 1965. Many at NASA believed the agency would remove an astronaut from a flight if his wife was pregnant, so Lovell and Marilyn had to figure out what to do. To him, the answer was simple—silence. He would keep training and say nothing; she would angle to be photographed from the neck up. By the time it became obvious that Marilyn was carrying, NASA would—hopefully—think it too late to change crews.

Training consumed all the astronauts' lives. By now, some of them were struggling in their marriages; the demands of the job, and the easy availability of women on the road, put a strain on their relationships. For Lovell and Borman it was different. Neither man caroused or stayed out late—not just because it wasn't the right thing to do, but because neither had the impulse to do it. They were in love with their wives—their best friends— women who'd loved them since the days when they were nothing but dreams, their lives just a blur of military base transfers.

At home in Houston, a very round Marilyn watched on television, eight months pregnant, as the Gemini 7 countdown neared zero. She didn't worry—she trusted in NASA and her Episcopal faith, and she trusted in Jim. When he'd left for the Cape, he hadn't given her any *if-I-don't-come-home* speeches or recited any *I've loved you forever* goodbyes. Instead, he swept the garage, balanced the checkbook, and painted the cradle in case the baby came while he was in space.

As photographers snapped her photo, Marilyn watched as Gemini 7's Titan II rocket spewed billows of orange-tinted smoke, then rose on a narrow, nearly transparent column of flame into the sky. The moment Lovell had waited for since Juneau High School was unfolding in thundering detail. It took seven full seconds before he could no longer contain himself.

"We're on our way, Frank!" he shouted to his crewmate.

At the two-minute mark, the spacecraft reached a speed of 3,600 miles per hour. Until now, the liquid-fuel rocket had lifted them in a kind of slow pull, but now the second stage kicked in, hurtling the ship forward with a new kind of fury. A minute later, Lovell and Borman were traveling at 7,100 miles per hour and picking up speed fast. Just under five minutes into the flight, Lovell caught a glimpse of something outside his window.

"Look at the Moon, Frank!"

The rocket pushed past seven g's and then separated from the spacecraft, sending Gemini 7 into orbit around Earth. For Lovell, the ascent was a wild and wonderful ride.

For the next several days, Lovell and Borman flew their spacecraft, conducted medical experiments, and, perhaps most astonishing for two men confined to such a tiny capsule, didn't drive each other crazy.

Toward the end of its two-week flight, Gemini 7 experienced

problems. The craft's fuel cells began failing and its thrusters faltered. Two days remained in the mission, and Borman's instinct was to terminate early. But Lovell—privately, without broadcasting a word for the public to hear—urged him to hang in and not worry, that the ship would make it. Along with Chris Kraft's reassurance, Lovell's encouragement persuaded Borman to hold on, and the flight finished near perfectly.

By the time the astronauts were aboard the aircraft carrier USS *Wasp*, they'd set several world records for space flight, including longest duration. Standing on deck, the scruffy Lovell said of the two cramped weeks spent with Borman, "We'd like to announce our engagement."

Back home, Marilyn told reporters, "Jim could come home beard and all, and I would welcome him with open arms." A month later, in January 1966, Marilyn gave birth to the couple's fourth child, Jeffrey.

Less than a year later, on November 11, 1966, Lovell was back on the launchpad as commander of Gemini 12. It was to be the final mission of Project Gemini. Strapped in beside him was Buzz Aldrin, who'd been selected as part of NASA's third group of astronauts in 1963. Together, the men would spend four days in orbit around Earth.

In some ways, the pressure on Lovell for this flight was even greater than it had been during Gemini 7. Gemini 12 had to succeed in order for NASA—and the country—to feel confident about launching Apollo, the program that would take America to the Moon. The mission went smoothly, and after a journey of 1.6 million miles, Gemini 12 splashed down in the western Atlantic. As Lovell was hoisted from the ocean by helicopter, he held a distinction that even he couldn't have imagined twenty years earlier, when he was writing letters to rocket societies and wishing he could afford college. Jim Lovell had now spent more time in space—eighteen days—than any other man in history.

Chapter Eight

●

PUSHED TO SUPERHUMAN SPEEDS

JUST TWO MONTHS REMAINED UNTIL THE SCHEDULED LIFT-OFF of Apollo 8, and even though the Apollo 7 flight had been a success, NASA still hadn't made the decision to green-light the mission. George Low, Chris Kraft, and others wanted to approve it immediately. George Mueller (pronounced "Miller"), Associate Administrator for Manned Space Flight, still had doubts, and remained deeply concerned about the risks and dangers inherent in Apollo 8's mission. That annoyed Kraft, who believed Mueller was being obstructionist because the idea to send Apollo 8 to the Moon hadn't been his. Still, Kraft remembered that Mueller had been willing to take risks in the past, ones that had paid off big for NASA. And Kraft couldn't disagree that the mission was risky.

As October came to a close, Mueller wasn't the only one concerned about the dangers of Apollo 8. By now, the media was openly discussing NASA's plan to send Borman, Lovell, and Anders to the Moon. "As the men in the space program [consider] Apollo 8," argued *The Washington Post,* "they must not allow anyone's desire to beat the Russians, or to get around the Moon by the end of 1968, or to fan public interest in the future of space exploration to enter into their calculations."

By this time, the astronauts were six weeks into their training with the command module simulators. To aid his memory, Anders had affixed small Velcro nameplates to several of the hundreds of switches, dials, and levers, little cheats that reminded him what was what. "Goddammit, Anders, it looks like a bunch of mayflies mating in here!" Borman said one day at the sight of all those plates. Anders used them anyway.

Before long, simulations advanced to incorporate various phases of the mission, all designed with problems built in. Some could be expected to occur on a complex mission like Apollo 8; others seemed million-to-one shots. Many of the most challenging scenarios came during highly critical parts of the mission, such as launch, exiting lunar orbit, and reentry, and those were practiced with particular intensity. But every part was worked through—over and over again.

But no matter how much practice and simulation, no matter how ingenious the scenarios run by the SimSups, astronauts could not be trained for everything. During the highly successful Gemini program, three of the ten manned flights had nearly ended in astronaut fatalities. And even if simulator training could cover the most complex and unlikely scenarios, the most basic malfunctions still could kill men in space. So Borman, Lovell, and Anders just practiced more.

As the weeks passed, the Apollo 8 astronauts got to know not just the spacecraft and the mission, but also one another. Borman and Lovell already knew they flew well together. Anders was the newcomer—and a revelation to the other two men.

In his six years at NASA, Borman had never seen a harder

worker, or a man of deeper integrity, than Anders. It was true that Anders had his own ideas about what was important, and didn't always agree with Borman on mission priorities, but he never went around Borman's authority or took a shortcut to anything. To Borman, character and competence counted more than most anything else, and Anders had plenty of both. Borman could think of no other astronaut in the entire program, longtime veterans included, he would have chosen over Anders for systems engineer.

Lovell, too, thought the team lucky to have Anders, and for many of the same reasons. He admired the way Anders had handled his initial disappointment with Apollo 8's change in assignment, one that likely meant he'd never set foot on the Moon. And he appreciated that Anders saw adventure and exploration in the chance to make man's first lunar journey. To both of them, going to the Moon wasn't just about beating the Soviets. It was a chance to do something incredible.

For his part, Anders felt welcomed by this old NASA duo. Like many, he considered Lovell a hail-fellow-well-met, just the kind of easy hand you'd want along on a six-day trip, whether to a fishing hole or to the Moon. Borman was another matter. Anders saw much of himself in the commander—the all-business demeanor, the intensity of approach, the swiftness and certainty of opinion. Sometimes when Borman barked an order at Anders, it was as if Anders was hearing it from himself. But that didn't mean he always had to sit there and take it, even if Borman outranked him.

"Look, Frank," Anders said one day, "my job is to make sure this spacecraft works, and I guarantee you that I'm going to know whether it's going to work or not. So you spend your time worrying about the mission and the rocket, and I'll worry about the spacecraft."

Borman respected that. *Character and competence.* The crew only got better after that.

Despite the synergy and good teamwork, it seemed to Lovell

and Anders that Borman might be dealing with a private stress, one not shared by his crewmates. Borman had a short fuse when planners tried to add superfluous tasks to Apollo 8; he angrily rejected NASA's idea of opening the hatch in space and adding a spacewalk to the flight. More than anything, Borman seemed willing to die on small hills—a rejection of NASA's new food, a refusal to allow a TV camera on board the spacecraft. Fighting these battles, Borman argued, was necessary in order to ensure focus on the flight's basic mission: Get to the Moon, orbit, get your ass home, beat the Russians, win. Add-ons and changes represented additional risk, and Borman wanted none of that.

It wasn't that Borman was wrong; on almost every one of these issues, he was right, and when he wasn't, as with bringing a TV camera so that the world could witness parts of the historic mission live, he eventually heard the good sense in others' arguments and relented. But Lovell questioned whether there might be an additional dimension to Borman's near-religious aversion to risk, and he couldn't help but wonder whether it might have something to do with Susan. Borman hadn't said anything about it, but Lovell had heard mention from others that Susan was terrified by the idea of Apollo 8's new mission, and that the memories of the Apollo 1 fire still burned in her mind.

As November rolled in, Kraft found himself facing a new problem: Even if Apollo 8 went perfectly, there would be no one in the Pacific to pick up the crew after splashdown, since the Navy's Pacific Fleet had already been given a reprieve for Christmas. Someone had to appeal directly to Admiral John McCain, commander in chief of the Pacific Fleet, to ask for special dispensation. The timing wasn't ideal; McCain's son, John McCain III, a Navy pilot, had been shot down over Hanoi and was being held as a prisoner of war by the North Vietnamese. But Kraft said he'd do it, and in person.

A few days later, he walked into an amphitheater-style conference room in Honolulu, surrounded by a hundred military captains, admirals, and four-star generals. At 10:30 A.M. sharp, an order sounded—*Attention!*—and McCain entered the room.

"Okay, young man," the fifty-seven-year-old McCain growled at the forty-four-year-old Kraft. "What have you got to say?"

Kraft described Apollo 8's mission, its benefits and risks, and explained that America's greatness was about to be tested in space. Then he laid out NASA's request. This part he'd rehearsed and memorized down to the word.

"Admiral, I realize that the Navy has made its Christmas plans and I'm asking you to change them. I'm here to request that the Navy support us and have ships out there before we launch and through Christmas. We need you."

For several moments, there was silence in the room. Finally, McCain got up from the table and slammed down the supporting documents Kraft had provided.

"Best damn briefing I've ever had. Give that young man anything he wants."

And with that, the Navy's aircraft carriers belonged to NASA for Christmas.

In Washington, Mueller continued to worry about Apollo 8. In a November 4 letter to one of NASA's top managers, he wrote, "you and I know that if failure comes, the reaction will be that anyone should have known better than to undertake such a trip at this point in time." Mueller also asked the man to fill out a Mission Risk Assessment Form. To Kraft, that was a portent of things to come—Mueller intended to make him and other top managers at NASA sign in blood that Apollo 8 was the right thing to do, and that they would be responsible if things went wrong.

On November 5, 1968, the American people elected Richard Nixon as the country's next president. During his campaign, Nixon had promised to support the space program, as Johnson had done. "I don't want the Soviet Union or any other nation to be ahead of the United States," he'd told voters a few weeks before the election. "Let's emphasize the Moon shot and others where we can make a direct breakthrough."

NASA managers continued to debate the Apollo 8 mission into November. As they went from meeting to meeting, an unmanned Soviet spacecraft lifted off from the launchpad in Kazakhstan. Zond 6 represented the final piece of the Soviet plan to send a crew on a circumlunar mission. Two months earlier, Zond 5 had made a successful loop around the Moon, only to experience a violent reentry that might have injured or even killed a crew. This time, Zond 6 had been designed to loop around the Moon, then execute a complex, guided reentry into Earth's atmosphere, reducing g-force loads to manageable levels. If the Soviets could pull that off, the next Zond flight would go to the Moon with two cosmonauts in early December—and beat out Apollo 8.

On November 11, NASA chief Thomas Paine made a final decision on Apollo 8. He phoned President Johnson, who was meeting with President-elect Nixon, and informed the men of the agency's decision. It was determined then that Paine would announce NASA's verdict on Apollo 8 to the American public at a press conference from NASA headquarters in Washington, D.C.

Early the next day, as Zond 6 headed on a perfect course for the Moon, Paine spoke to members of the media. "After a careful and thorough investigation of all the systems and risks involved," he said, "we have concluded that we are now ready to fly the most advanced mission for our Apollo 8 launch in December, the orbit around the Moon."

The press conference lasted more than three hours. When it ended, reporters rushed their stories to their respective outlets. America was shooting for the Moon at Christmas.

Three days after Paine's announcement, a letter arrived at the office of Bob Gilruth, director of NASA's Manned Spacecraft Center in Houston. It was written by a man named Stewart Atkinson, of Darien, Connecticut. It read:

Dear Sir:

I wonder what sort of thinking went into your decision to send three men around the Moon at Christmastime. This is by no means a sure venture, and the risk of ruining the Christmas Season for millions of Americans is enormous. Christmas is a time for carefree family reunions, for as much happiness as all of us can snatch in this miserable year of 1968. We do not need a space triumph to celebrate our greatest holiday, but a failure will be the crowning blow to a people already punch drunk with the events of the year.

Along with millions of Americans I have been thrilled by the successes of the Space Program . . . but I am of the opinion that the American people would much prefer a delay of a month if such is essential.

Sincerely,
Stewart Atkinson

It was around this time, about six weeks before scheduled liftoff, that Borman got a call from the agency's public affairs mastermind, Julian Scheer. NASA, Scheer said, had decided to have the crew of Apollo 8 make a live television broadcast on Christmas Eve.

"We figure more people will be listening to your voice than that of any man in history," Scheer said. "So we want you to say something appropriate. You'll have maybe five or six minutes."

"Great, Julian," Borman replied. "What are we doing?"

"Do whatever's appropriate," Scheer said.

Borman was surprised by the response. Scheer, and NASA,

were leaving it up to him to decide what to say. No committees. No consensus. No vetting. Just him.

By this time, the unmanned Zond 6 had already flown around the Moon, passing within 1,500 miles of its surface, and was headed back to home. So confident in the mission were Soviet planners that they took the uncharacteristic step of announcing, during the flight, that the explicit purpose of the mission was to prepare for a manned journey to the Moon. All that remained was for the spacecraft to execute its complex reentry and touch down under parachute in Kazakhstan.

Execution was near flawless. Zond 6 completed its reentry having endured no more than four to seven g's. The flight of Zond 6 made it clear to NASA that the Soviets were ready to send men to the Moon ahead of Apollo 8. And the Soviets didn't intend to stop there. One of their experts said that the flight of Zond 6 paved the way for manned flights not just to the Moon but to Mars, Venus, and other planets.

What NASA, and even the CIA, did not know was that Zond 6 had experienced two serious problems during its flight. The first, a partial depressurization of the cabin, occurred just before reentry. The second, a failure of the parachute system, caused the spacecraft to plummet into the ground. Both incidents would have been fatal had a crew been on board.

That meant the Soviets had a decision to make. Given the problems with Zond 6, should they risk sending a crew to the Moon aboard Zond 7 in early December? Or should they make one more unmanned lunar flight to make certain those problems had been worked out? Those who wanted to go, including the cosmonauts, felt certain the problems on Zond 6 could be fixed, and were willing to take their chances. Those who preferred to play it safe couldn't stand the thought of losing another cosmonaut in flight, as they had in a 1967 accident that still haunted the country. And many of them didn't believe the Americans crazy enough, in any case, to launch Apollo 8 in December. The Soviets had already sent two flights capable of carrying men to the Moon; the Americans had sent none. NASA, they figured, would soon come to its senses and order Apollo 8 to stand down.

In Houston, many worried that NASA might decide the same. Nervous personnel counted down the number of days until the next Soviet lunar launch window opened. In Kazakhstan, the Soviets moved a new Zond spacecraft to the launchpad, a ship scheduled to lift off for the Moon two weeks before Apollo 8.

Few in America knew that this spacecraft even existed. All that anyone knew in the West was that the race to the Moon was being pushed to superhuman speeds that could get men killed. A recent newspaper editorial had proposed cooperation between the United States and Soviet Union, a fixing of the race so that the two nations arrived simultaneously, a way to avoid a tragedy. But neither side had come this far to compromise. And neither had Apollo 8's youngest astronaut, a thirty-five-year-old father of five who'd been born in Hong Kong and came from fighting stock, the kind that would never settle for a tie.

Chapter Nine

●

BILL ANDERS

WILLIAM ALISON ANDERS FIRST WITNESSED AN ATTACK FROM the sky by foreign invaders in 1937, when he was four years old. He was living with his parents along the Yangtze River in Nanking, China, when his father, a United States Navy lieutenant, sensed that Japanese forces would attack nearby Chinese boats. Arthur Anders told his wife, Muriel, to take their son and evacuate. After a two-day trip by train to Canton, mother and son found a hotel room, and it was there that Bill watched Japanese airplanes streak overhead and bomb ships in the Pearl River just two hundred yards away.

The next day, Bill and his mother boarded a boat and made their escape. Bill's father stayed, manning the American gunboat USS *Panay*, on which he was second in command. A few days later, on December 12, 1937, Japanese aircraft attacked the *Panay* as it moved up the Yangtze. The United States was a neutral party in the conflict between Japan and China, and the boat, marked by American flags, was attempting to move people to safety. A bomb struck the boat's bridge, wounding and disabling the captain. That left Arthur Anders in charge.

Despite America's neutral standing, he ordered the *Panay* to open fire on the attacking aircraft. Badly injured in sickbay, the boat's captain wanted the crew to abandon ship, but Anders wouldn't have it. "He's not in charge anymore, I am," Anders said. The *Panay* was not outfitted to engage attacking aircraft, but Anders directed the fight nonetheless. Soon dive-bombers appeared from the smoky skies, unleashing a second attack on the damaged American boat. Still Anders ordered the crew to continue to defend, even as the *Panay* slowly began to sink. Realizing that crew had been injured, Anders attempted to man one of the boat's guns himself, taking shrapnel wounds to his hands.

As Anders stood on the bridge, a piece of shrapnel pierced his throat, causing heavy bleeding and making it impossible for him to speak. Using his own blood as ink, Anders scrawled out directions to the crew on a chart, and the fight continued. Eighty minutes after the attack started, desperate men made their way to small escape craft. Anders was last off the boat and then lost consciousness. By the end, two Americans and an Italian journalist from the *Panay* had died, dozens had been wounded, and the sinking became an international incident. Realizing that it had committed an act of war against the United States, Japan apologized.

Arthur Anders received the Navy Cross, the highest honor bestowed by that branch for a peacetime action. The orders he wrote in blood are preserved in the National Museum of the U.S. Navy in Washington. The prelude to the fight would remain one of young Bill's earliest memories.

Bill's parents, both Americans, had met in the Philippines during Arthur's tour of duty there.

Muriel's father was the civilian in charge of the Cavite Navy Yard, which repaired American ships. Bill, the couple's only child, was born in Hong Kong. The family moved often when Bill was young, eventually returning to America in 1938. Through his childhood, Bill absorbed the Navy life, and he expected to attend the Naval Academy, as his father had.

When he was fourteen, Bill moved with his family to Weimar, Texas. As Arthur drove Bill to school one day, father and son spotted a biplane in a field, along with a banner hanging from a fence: AIRPLANE RIDES—FIVE DOLLARS.

A few minutes later, Bill and the pilot were soaring over open fields.

"Want to do a loop?" the man asked.

Bill nodded.

The pilot was low for that kind of maneuver, no higher than two thousand feet, but he pulled up, looped over, and managed to just miss the ground as he righted the plane.

Bill had a hard time concentrating in school that day; no matter how hard he tried to focus, his mind kept looping over Texas.

Driving home that afternoon, Bill and his father came upon the field where Bill had flown. The plane was still there, but this time it was jammed nose down into the ground, a terrible crash. When Arthur inquired, he was told two people had been killed during a ride. Bill looked at the seat he'd occupied in the now-fractured craft, and remembered how close he'd come to the ground on his loop. On airplanes, it seemed, the difference between life and death could come down to a few feet.

Bill began high school in Texas, but he moved with his family to the San Diego area to begin his sophomore year. By then Arthur had been made a Navy reservist as a result of his wartime injuries and was working at the naval training station. He and Bill played catch, took car rides, and went on San Diego Mineral and Gem Society trips; as a boy, Bill had fallen in love with natural history and geology, and he resolved to own a piece of every kind of rock in the world. Sometimes, the men would go high into the Sierras looking for specimens; it was on trips like that when Bill noticed that he was willing to travel almost anywhere as long as there was something new to find.

Bill became president of his high school's biology club, largely on the strength of his expertise on snakes. He read books, many on science, often finishing them in one day. Instead of science fiction, Bill preferred to read about old ships from bygone eras, and about pirates and life on the high seas. Those were men who'd undertaken real adventure, who'd pushed themselves into actual, not theoretical, unknowns.

As the second-smallest student in his class, Bill found it hard to make time with the ladies (his love of science and snakes didn't help). Despite her son's size, Muriel encouraged Bill to play football. He suited up and was knocked flat, but he loved the feeling of getting up and realizing he had survived.

After his junior year of high school, in 1950, Bill transferred to a military prep school in San Diego. Since early boyhood, he'd envisioned a life like his father's—defending his country on board a ship, fighting back. Military school would give him the best chance for admission to one of the nation's service academies. In a different time, one in which America had lesser enemies, Bill might have become a geologist. Now, in the teeth of the Cold War, he headed for Annapolis.

Bill Anders arrived at the United States Naval Academy in 1951 with the ambitious goal of becoming an officer aboard a destroyer and making four-stripe captain. By Christmas, he was about to wash out. He'd skated through high school on brains alone, but that level of effort wasn't cutting it at the Academy, even as the son of a Navy Cross recipient. An adviser warned him he wasn't long for the place unless things changed. Anders straightened up.

Having survived his first year at the Academy, Anders returned to San Diego for the summer. There he found himself on a double date at the beach, but when he saw his friend's date, he forgot about his own. Sixteen-year-old Valerie Hoard was about the prettiest young lady Anders had ever seen, and she had a quiet confidence beyond her age. Anders spent the day swimming alongside Valerie as she lounged on an inflatable raft, asking about her life, hearing her descriptions of how her father gave her rides on the back of his California Highway Patrol motorcycle (sirens blaring and red lights flashing). Anders never stopped to catch his breath as they toured all over Mission Bay. *This guy has a lot of endurance,* Valerie thought. When Anders finally dropped her off at home, he shook her hand and said goodbye.

Summer was drawing to a close, so Anders had to make the time count if he hoped to keep seeing Valerie. On their first official date, he took her to the Navy officers' club, and then to the Starlight Bowl to see the San Diego Civic Light Opera. The next night, he took her to the Old Globe Theater for Shakespeare. It was heady stuff for Valerie, and she was impressed with this serious young man. At home, she asked her mother why Bill shook her hand after dates but didn't kiss her. The truth was that Anders didn't have much experience with girls and didn't want to push his luck. That was fine with Valerie—she had other suitors to keep her company. A few days later, Anders was back at the Naval Academy, and Valerie was back in high school.

By his second year at Annapolis, Anders was rising up the class rank. He always found time to write letters to Valerie, long ones, every day, about his outlook on life, the challenges of the Academy, how he saw the world. At Christmas, when he was home, they spent every day together. Not once since the day he met her had Bill doubted that Valerie was the one for him. She was poised and gracious, self-assured even in unfamiliar situations, and seemed curious about everything. She was a popular and busy senior who hardly had time for serious romance, yet she was slowly falling in love with Anders, and he was in love with her.

The relationship did not please Muriel. She'd long thought

her son should marry an admiral's daughter—a higher grade of folk—and took the formal tea dance invitations he'd received and lined them up on her kitchen window. Valerie saw the display when she was at Anders's house, but she also noticed something else—that he hadn't attended a single one of these debutante parties. He just wanted to be with her.

In the summer before Anders's third year at the Academy, he and about four hundred classmates boarded the USS *Bennington,* an aircraft carrier bound from the East Coast for Halifax, to see how fliers operated at sea. Also aboard was an array of fighter aircraft: Panthers, Cougars, Crusaders, and the AJ Savage, a three-engine nuclear-weapon-carrying bomber. On the first night, a young Marine pilot made a landing approach in his Cougar, floated over all the wires, and slammed into a pack of parked airplanes. Such was the surplus of aircraft after the Korean War that sailors just pushed the damaged ones overboard rather than fix them.

Hours later, an AJ Savage came roaring in and hit badly on landing. The pilot and copilot tumbled down the flight deck head over heels in their severed, flaming cockpit but somehow managed to survive; the third crewman, however, died when he was thrown under the ship.

The smoke had hardly cleared on that incident when Anders saw one of the gull wings of a Corsair fold up during takeoff. Just off the flight deck, the plane did a full roll and plummeted into the water.

Immediately, the carrier turned toward the downed aircraft to make a rescue. Anders could see the pilot in the cockpit, but it was clear the man wasn't moving. Anders had been on the plebe swim team and could handle himself in rough waters; now he had a decision to make. He could jump in and try to rescue the pilot, or he could allow carrier rescue personnel to do what they were trained to do. The sight of the pilot, unresponsive and starting to sink, pulled on him, but he also knew the ship was moving at about thirty-five knots, he had no life jacket, and he'd have to fall about fifty feet before hitting the water. He had a thought that would bother him for years: If he did jump, he might get put on report or receive demerits. He saw a helicopter and a destroyer approaching to assist in the rescue, and in a split second he made his decision to stay aboard the ship. Rescuers couldn't reach the scene, however, before the pilot and his airplane disappeared under the waves.

Anders hardly knew what to make of the disasters he'd seen. Navy pilots were trained to be the best in the world in combat, yet they risked their lives every day, even during takeoff and landing. Still, an airplane had the power to take the fight to an enemy with an immediacy unavailable to giant ships. It was more personal, too, just pilot and machine as one. When it came time to decide what to do with his military career, Anders wanted nothing to do with aircraft carriers, but knew he had to fly.

Anders continued to write to Valerie every day. Despite worries that she would turn him down, he bought an engagement ring and invited her to the Naval Academy's formal Ring Dance, at which couples would dance through a replica of the cadets' class ring. Valerie and Bill held each other close as they moved around the dance floor to the sounds of a big band. Valerie wore Bill's class ring on a chain around her neck and the engagement ring on her hand.

Valerie still wasn't quite eighteen. Marriage meant giving up a college education, which was important to her. It also meant making a life with a man who'd chosen a dangerous line of work. But her father chased bad guys on his motorcycle for a living, and twice he had almost been killed on the job in accidents, so she was used to living with risk.

There was also the matter of religion. Anders's father was a strict Catholic, and his church would insist that Valerie be Catholic, too. In the end, that also seemed fine to a girl in love, and even though she was still in high school, Valerie said yes, knowing that Navy rules didn't allow midshipmen to be married until graduation, so a yes for the future—not tomorrow, but a yes nonetheless.

As a high-ranking member of his class, Anders had options with his career. He knew he wanted to fly, and he could decide between a commission in the Navy or the newly formed Air Force (established just eight years earlier). Choosing the Navy meant operating from short carrier decks. Choosing the Air Force meant flying from ten-thousand-foot concrete runways. Anders chose the Air Force.

Shortly after graduating from the Naval Academy in 1955, Anders married Valerie in a Catholic ceremony at the naval chapel in San Diego. He then reported to Air Force flight training near the Rio Grande Valley in Texas, where he began flying the T-34 Mentor. By the next stage in training, in the much bigger T-28, he realized that he had a natural ability. Sometimes he'd invite Valerie out to a dusty crossroads and put on a private airshow for her, flying too low, testing to see how much vertical pull-up he could endure before blacking out from loss of oxygen to the brain caused by high g-forces, seeing if he could wake up before the plane went down. Valerie loved her husband's performances. She also liked that he didn't play things exactly by the book, that he took risks. To Valerie, the most interesting lives often seemed to go that way.

After earning his wings at age twenty-three, Anders was assigned to an Air Defense Command all-weather interceptor squadron at Hamilton Air Force Base near San Francisco, where he would fly the twin afterburner F-89 Scorpion. Interceptors flew to prevent enemy aircraft from penetrating restricted airspace, either by chasing them off or by engaging them in combat. One model of Anders's jet was armed with two rocket-propelled missiles, each with a 3.5-kiloton nuclear warhead attached— combined, it equaled about half the explosive power of the bomb dropped on Hiroshima in 1945. To fire the weapons, the radar operator in the backseat had to throw a switch, and the pilot in the front seat had to throw his own switch. That's all it required. Officially, the crew needed an order from the ground, but if Anders and a buddy wanted to start World War III, they could do it on their own. "That's the Cold War," Anders told Valerie. "It's up to us not to screw up."

In February 1957, the Anders family welcomed their first child, Alan. And in July 1958, Valerie gave birth to Glen. Raising a young family in California was idyllic, with the warm weather and abundant culture, maybe too good to be true, so it came as little surprise when Anders got a new assignment: Iceland.

Valerie would stay with the kids in California while her husband moved four thousand miles away. Again, Anders's job was to fly interceptors. This time, he would be going after Soviet bombers, long-range machines that flew missions near Iceland and the North Atlantic designed to test American air defenses. To help avoid starting a world war, his aircraft and others would be armed only with conventional air-to-air rockets, no nukes.

Early in his assignment, a Soviet bomber penetrated the eastern edge of Iceland's Air Defense Identification Zone. Anders and his wingman scrambled into the air, afterburners blazing, and caught up with the Russian plane. Anders positioned his wingman to shoot down the bomber if its pilot gave the Americans any trouble, then flew his F-89 so close he could call out the eye colors of the Soviet crew. The Russians smiled and waved. Anders offered his own American greeting—a middle finger.

The Soviet crew kept smiling and waving, then broke back to where they belonged.

Low on fuel, Anders returned to base, knowing the incident would be important to American intelligence officers, as it was among the first—if not the very first—intercept of a Soviet bomber in the zone. On the ground, he described the event.

"Anything else?" asked a representative of the Defense Intelligence Agency.

Anders feared he would be facing some discipline. Still, he had to be honest.

"There is something else," Anders said nervously. "I probably should tell you that, you know . . . I gave them the finger."

The man smiled. There was no trouble.

Anders flew more missions in Iceland, many of them risky, both for the dangerous flying conditions and for the potential conflict with Soviet bombers. Three or four months after Anders flipped off the Russian crew, another pilot in his squadron intercepted a Soviet bomber. This time, the Russians had a response to their American pursuers, and they held it up to their window—a sign printed in English—for the Air Force pilots to see.

American intel had a good laugh when they heard the story. To them, it represented the layers of bureaucracy that constituted the Soviet socialist system. It had taken more than one hundred days for the first bomber crew to report Anders's middle finger, for word to travel through channels to the Kremlin, for analysts to decipher it, for committees to formulate a response, for other committees to approve it, for translators to put the Soviet answer into English, for orders to be given to a new bomber crew, and for the Soviet pilots to deliver it.

Their message to the Americans flying alongside: WE FUCKED YOUR SISTER.

After more than a year in Iceland, Anders was sent back to Hamilton Field in California, a welcome return for Valerie. Anders continued flying interceptor missions, this time with the nuclear-armed supersonic F-101 Voodoo, a fearsome jet capable of reaching speeds in excess of a thousand miles per hour.

At Hamilton, Valerie became even more accustomed to the stresses of being married to a fighter pilot. Men died in this line of work, she knew that, but it was always terrible to see a black Air Force car drive into base housing to deliver the bad news. Every time she saw the black car she wondered, *Is my life about to change? Could this happen to us?* And even as the car passed her home and stopped at a neighbor's, she didn't kid herself. *Yes,* she thought, *it could certainly happen to us.*

Around this time, Anders began to get itchy. Interceptor

work was interesting, but he didn't feel pushed to his limits, not in body or mind, in a way that would make for a satisfying long-term career. He went to see Chuck Yeager at the Test Pilot School at Edwards Air Force Base in Southern California. Pushing unproven airplanes to their limits demanded a new level of intellectual engagement, raw bravery, and adventure; to Anders, that sounded like the life he wanted.

Yeager was impressed by Anders's flying credentials but urged him to go back to college and obtain an advanced degree in science or engineering, since that's what the Air Force was looking for in test pilot candidates. Anders followed the recommendation and applied to the Air Force Institute of Technology at Wright-Patterson Air Force Base near Dayton, Ohio. He requested a program in either aeronautical or astronautical engineering, but administrators put him in nuclear engineering. To cover his bases, Anders enrolled in a night school program in aeronautics at nearby Ohio State University.

Over the next two years, Anders studied, fathered another child, Gayle (born 1960), and learned more about nuclear energy and radiation. In 1962, he graduated second in his class with a master's degree in nuclear engineering. He submitted his application for test pilot school, but now the school wasn't accepting new students. Dejected, he chose to go to the Air Force Special Weapons Center in Albuquerque, New Mexico, to work on a radiation shielding project and instruct pilots in jet aircraft. While in Albuquerque, Valerie gave birth to the family's fourth child, Gregory. All the while, Bill waited for an opening at the test pilot school. Valerie took an astronomy course at the University of New Mexico, just out of fascination with the subject, a baby on her hip.

In June 1963, Anders was driving in his Volkswagen Microbus when he heard a news broadcast on the radio. The announcer said that NASA had decided to add a third group of astronauts. Anders met every one of the agency's requirements: age limit thirty-five, two thousand hours flying time in advanced jets,

maximum height six feet. "Must also be a test pilot," the man
said. Anders's heart sank. "Or the applicant must possess an advanced degree." Anders wondered if he'd heard the last part correctly. He pulled over to the side of the road and waited, through
twelve minutes of commercials and bad music, for the next newscast. He had heard correctly—one needn't be a test pilot to
apply. He wrote down NASA's address. He'd been interested in
astronauts since the Mercury 7, the United States' first group of
astronauts, had arrived on the scene four years earlier, but space
travel had never seemed possible for mere fighter pilots. Now,
things had changed.

It would be his dream job in many ways. Joining NASA would
give Anders the intellectual stimulation he craved, the chance to
fly the most advanced machines ever built, and the opportunity
to become an explorer, a space-age version of Charles Lindbergh
or Vasco da Gama, the New World voyagers he'd always admired.
And he could bring back unknown rocks from his journeys to
the Moon.

And there was another benefit, one that resonated with a man
whose father had fought back against America's attackers, even
when the United States wasn't formally at war: He could do more
in space than anywhere else to help defeat the Soviet Union.

That night, Anders drafted a letter to NASA describing his
qualifications: world's greatest pilot; can solve all space radiation
problems; jet instructor; great guy. Valerie typed version after
version. They sent the final copy, by certified mail, the next day.
It arrived with four thousand other letters penned by astronaut
hopefuls.

To Anders's amazement, he was asked to report, along with
about a hundred others, to Brooks Air Force Base in San Anto-
nio, Texas, for a physical. There, he was put through a battery of
tests, not just physical but psychological.

Near the end of the process, only twenty-eight finalists remained. Anders had to appear before the so-called Murder
Board, a group of final interviewers that included current astro-

nauts, Chris Kraft, and a doctor. He had little trouble with the questions from the space people. The doctor was another matter.

"Well, Captain Anders," the man said, "your record looks pretty good. But we're worried about this concussion you had in the past."

Anders had never suffered a concussion. Could the doctor be trying to trip him up? Test him? Or maybe the doctor had another applicant's records and believed he was interviewing a different candidate.

Anders's mother had taught him never to lie. But she'd also reminded him that he needn't always blurt out the full truth, either. On the spot, he formulated an answer.

"Sir, I've never been bothered by a concussion."

"Bothered" was the key word. *That* was true.

On his thirtieth birthday—October 17, 1963—the phone rang in the Anders home. Valerie handed him the receiver. It was Deke Slayton calling with a job offer. Anders never did figure out if the doctors had been looking at the wrong guy's records. And as Slayton offered him a job, he was much too happy to care.

NASA assigned each of its new astronauts to a specialty. Anders focused on radiation and environmental controls—cabin pressure, temperature, carbon dioxide, and so on.

He also focused on potholes. After complaining about the condition of the roads near his new house, the town council named him street commissioner, a job he would hold, concurrent with his job as astronaut, for the next two years.

Early in training, Anders gravitated toward two of his fellow new astronauts, Walt Cunningham and Rusty Schweickart. All three men had an intellectual bent, and all three were interested in space science. Not one had been a test pilot. Together, the trio tackled the single most vexing question at NASA: How does a new astronaut best position himself to get selected as soon as possible for a space flight? After careful analysis, they determined to increase their physical fitness, become more expert in their specialties, and further master the science of space travel.

None of it made a ripple. To Anders, it seemed the more he and his pals tried, the more invisible they became to Slayton, the man who assigned astronauts to flights.

And then it dawned on Anders. Slayton considered him, Cunningham, and Schweickart to be nerds. Slayton didn't seem to give a damn about Anders's advanced degree in nuclear engineering, or Cunningham's doctoral work in physics, or Schweickart's research on upper atmospheric physics at MIT. He certainly didn't seem to appreciate that Anders had signed up for extra geology field trips. Selection appeared to come down to two criteria: seniority and one's standing as a test pilot. And that wasn't good news for Anders or his friends.

It all struck Anders as unfair, but he still had to look for an edge. It seemed to him that Slayton, an avid hunter, liked astronauts who joined his hunts. Anders had little interest in shooting game, but when an invitation to an antelope hunt went out, he signed up. Slayton and at least a dozen astronauts packed rifles and flew to Lander, Wyoming. On arrival, each was given a single bullet; it was a one-shot hunt, and that's all the ammunition they were allowed. Anders wasn't going to shoot at an antelope unless he was certain he could hit it. And yet he knew he couldn't return to camp with an unfired bullet; nothing would cement the view of him as a square more than that.

After a time, he spotted an antelope walking peaceably a few hundred yards away. Anders had been on the Air Force pistol team and was a good shot. He hated to do it but aimed his rifle and fired. His bullet tore into the antelope's hindquarters, sending the wounded animal running and bleeding.

Anders followed the trail, then killed the antelope with his knife, all the while apologizing to the poor creature and thinking, "This is the last goddamn antelope hunt I'm going on." He knew astronauts were supposed to do manly things. But he also knew a hunt like this wasn't him. He determined never to go on another.

Back in Houston, another astronaut, Alan Bean, joined Anders, Cunningham, and Schweickart in their unofficial group. Bean had been a test pilot, but as an avid painter, he seemed more artist than warrior. By now, Anders should have realized it didn't pay to look like an egghead, but since he and his friends weren't being put on crews anyway, they decided (with the exception of Bean) to enroll at Rice University to pursue PhDs. For all Anders knew, he was destined to sit on the sidelines forever.

His fortunes, however, changed in early 1966, when he became CapCom for Gemini 8. (CapComs, or Capsule Communicators, were the astronauts at Mission Control who communicated by radio with the crew.) A few minutes after Anders came on duty, pilot Dave Scott radioed to Control from space.

"We have serious problems here. We're—we're tumbling end over end up here."

The spacecraft was rolling violently out of control while in orbit around Earth. Suddenly NASA was face-to-face with disaster. While Anders calmly relayed information from Mission Control and reassured the astronauts, commander Neil Armstrong battled to regain control of the ship by using the craft's reentry thrusters. Several agonizing minutes later, Gemini 8 had been steadied. The mission was terminated early and the crew survived. Anders didn't think he'd done anything special; he'd just stayed cool under pressure, and it was Scott and Armstrong who deserved credit for a terrific save. But after Gemini 8, Anders registered brighter on Slayton's radar.

Not long after, he was assigned, along with Armstrong, to be the backup crew for Gemini 11. He then joined Frank Borman's crew after the Apollo 1 fire in early 1967. Along with Mike Collins, he and Borman would man Apollo 9. (Owing to problems caused by a bony growth between his neck vertebrae, Collins would later be replaced by Jim Lovell.) The flight would be a high Earth orbit checkout of the full Apollo spacecraft. It wouldn't go to the Moon, but it would put Anders in position to make that journey—and to walk on the lunar surface—on a subsequent Apollo flight.

Anders spent long stretches away from home during training. Valerie was raising their five children (the family had welcomed another son, Eric, after Anders joined NASA) in El Lago, a small town near Houston where many astronauts lived, making Bill's paycheck go seven ways. One day, Anders calculated the amount of time he spent with each of his kids: eleven minutes per week per child. He regretted it, and didn't consider himself to be a good father because of all the time he spent away. But for now, beating the Russians was more important than being an ideal family man.

Valerie saw it much the same way. She would have preferred her husband to be home more often, but she believed in NASA's mission, and in winning the Space Race. Even if it wasn't easy running a household by herself, things never got boring for Valerie. She was interested in science and technology, and in astronomy, so she watched with special interest as America worked its way to the Moon, and she made sure her kids watched, too.

One day she took all five of them to Ellington Air Force Base to see Bill fly the Lunar Landing Research Vehicle, or LLRV, the closest thing engineers could build to approximate the lunar module astronauts would land on the Moon. It was the weirdest ship any of them had ever seen, with a gimbal-mounted J-35 jet engine, sixteen lift rockets, and seating for the pilot that looked like an outhouse without a door. Tubular arms and legs jutted in every direction. Together, the jet engine and lift rockets could simulate flying in one-sixth gravity—equal to that on the Moon. Anders and Neil Armstrong took turns as pilot, making it look as if they were descending to the lunar surface. Aboard the ship, Anders felt like NASA's golden boy, one of the few who'd already been chosen to land on the Moon.

When the flying ended, three-year-old Eric picked up a loose screw and swallowed it. NASA doctors took X rays—a child never received a more state-of-the-art examination. It was a day unlike any Valerie had experienced.

Twenty-four hours later, on May 6, 1968, a system failure caused Armstrong to lose control while he was flying the LLRV. At an altitude of less than two hundred feet, the machine pitched sideways and plummeted toward the ground. Armstrong ejected just moments before the craft impacted and burst into flames. Even with his parachute, his descent lasted only ten seconds. When Anders told Valerie the story, she didn't get upset or ask her husband to reconsider his mission. She thought, as she often did, *This is the life we've signed up for.*

In August 1968, Anders told Valerie that he was going to the Moon. Though the mission would be rushed, it seemed to her like an opportunity. She couldn't escape the feeling that America was in very bad shape. She'd seen the endless television coverage of race riots, Vietnam protests, and assassinations, and asked herself, *Where is our turning point? Where are we going to find hope?* As a stay-at-home mom with no help, Valerie was busy from morning to night, taking care of her five children, mowing the lawn, trying to make ends meet on her husband's military salary—she had little time to figure out how to save America. But she knew that the chance for Apollo 8 to rise to a near-impossible challenge could be a positive statement in the country's crushingly negative year.

Not long after Bill began training for Apollo 8, Valerie ran into George Low at a social event. They didn't speak of the new mission, but Valerie could see in Low's eyes that he had reservations about Apollo 8, that he'd contemplated what a huge step this was for NASA, that he knew how much could go wrong. She felt for him. *What a difficult position he must be in,* she thought, *to be responsible for all this.*

As the launch date for Apollo 8 grew nearer, Anders had even less than the usual eleven minutes per week to give to each of his children. On one rare day off, he woke his family early and took them to water-ski behind the tiny boat he owned, one with a 40-horsepower motor. In a few weeks, he would be riding an engine 4 million times more powerful than that, but for now, it was all the power he needed, enough to last for an entire day.

Chapter Ten

●

HOW'S FIFTY-FIFTY?

THANKSGIVING WAS JUST THREE DAYS AWAY, AND LESS THAN four weeks remained until the scheduled launch of Apollo 8. While most Americans got ready to celebrate the holiday, Borman, Lovell, and Anders were hard at work with the SimSup.

The focus during these pre-Thanksgiving sessions would be on two key aspects of the flight. The first, Lunar Orbit Insertion (LOI), would come when the spacecraft arrived at the Moon and fired its Service Propulsion System (SPS) engine in order to slow down enough to be captured by lunar gravity and go into orbit around the Moon. The second, Trans Earth Injection (TEI), would come when the spacecraft fired that same engine to pick up enough speed to leave lunar orbit and head back to Earth. Both of these critical maneuvers would occur around the far side of the Moon, completely out of touch with the engineers on Earth who might catch any equipment malfunctions or slip-ups by the crew. More than almost anything else, it was TEI that worried the astronauts, controllers, engineers, and NASA officials. The SPS engine had no backup—if it misfired or didn't fire at all, the spacecraft and crew could crash into the Moon, fly off into endless space, or be trapped in a slowly decaying lunar orbit that would ultimately impact the lunar surface.

The simulations began early in the morning. More than once the astronauts perished because someone didn't fix problems correctly or in time. In those cases, the crew and controllers held a short briefing afterward, discussed how they'd failed and what could be improved, then tried again. Over and over, scenarios were run, often for full days at a time, the more catastrophic the better, until repetition began to groove instinct into all the participants, and dying helped the men learn to survive.

Even at NASA, Thanksgiving was a day for family, and Borman, Lovell, and Anders found their way home just in time to celebrate the holiday. Apollo 8 was scheduled to launch in just twenty-three days, so Thursday was the only day off the men were allowed.

By now, all three wives had decided where they'd be when Apollo 8 lifted off. Marilyn Lovell wanted to see it live, to be as close to it and as much a part of it as possible. (She had been too far along in her pregnancy to watch Jim launch on Gemini 7, but she did witness his flight on Gemini 12 in person.) And she wanted her four children to be there, too; it was something she thought they should experience as a family. That was fine with Lovell; while he knew there was a chance of disaster, he never thought about launches, or life, in those terms, and he didn't want his children to think that way, either. A person had to take things as they came.

Lovell spread out his maps of the Moon for his children and showed them which parts of the lunar surface he'd be flying over and what the crew intended to do there. He'd even brought his children to explore the simulator at the Cape. He didn't tell them that his mission was dangerous, or that their father might not be coming home. There was no reason to put a fear like that into children.

Valerie Anders made a different decision. Her children were younger than Marilyn's, the flu was going around, and she didn't want to risk having five sick kids while holed up at a hotel near Cape Kennedy. The decision to stay home in Houston made a lot of sense to Anders. Even if his kids remained healthy, he wanted them and Valerie to be home, in a comfortable and safe environment, in case a disaster unfolded.

In any event, Anders got the sense that his kids were more interested in water-skiing and playing with their friends, which was fine with him. His eleven-year-old son, Alan, told how a classmate brought his fireman father to school one day, and everyone thought that kid had the coolest father of them all; in this neighborhood, it seemed every old dad was an astronaut.

When Valerie talked to the kids about Apollo 8, she explained what their father would be doing, but she never promised that he would be okay; she didn't want to mislead them. And they didn't seem worried, anyway. They had other excitements to deal with, like the new color television Anders bought so his family could watch the launch, and the *Life* magazine photographers who were now showing up almost daily to take photos of the family doing things they didn't do in real life, like eating ice cream together at the kitchen table.

When Borman arrived home for Thanksgiving, Susan had the house perfect and ready for him, as she always did. In the 1950s, when they were moving from base to base, she'd read and absorbed *The Army Wife* by Nancy Shea, a book about making a good life and a good home while married to a military man. "Every Army wife has three basic responsibilities," the author wrote:

1. To make a congenial home
2. To rear a family of which he will be proud
3. To strengthen her husband's morale

"Your whole scheme of life revolves around your husband, your children, and a happy home," Shea added. Susan had been the perfect Army wife for eighteen years, since Frank had graduated from West Point in 1950, but now she wondered whether she'd be able to keep it up, whether the pressure might finally break her.

Susan believed, with one hundred percent certainty, that Frank was going to die aboard Apollo 8. Frank knew she had been drinking, but he didn't think she had a problem because she kept such a beautiful home, raised her sons with honor and dignity, and never expressed a moment's concern for herself. He'd never seen her drunk, not once. When he discussed Apollo 8 with Susan, she told him, "I know you'll be fine." Inside, she was dying, watching her life and the lives of her sons being torn apart before her eyes, her best friend being taken away forever, to a place she could never reach.

If Borman had had even an inkling that his wife was suffering, he would have explained the mission to her, laid out maps, described how thoroughly NASA had engineered the flight, listed all the precautions that were in place. If her sons, now seventeen and fifteen, had known that their mother was so worried, they would have hugged and reassured her. But no one knew, which was just how Susan wanted it—she didn't want anyone to suffer on her account.

As launch neared, Susan hosted a cocktail party, with many NASA folks in attendance. During a quiet moment, she pulled Chris Kraft aside.

"Chris, I'd really appreciate it if you'd level with me," Susan said. "I really, really want to know what you think their chances are."

"You really mean that, don't you?" Kraft asked.

"Yes. And you know I do. I really want to know."

Kraft respected Susan. He thought she deserved an honest answer, and he did not want to sugarcoat things for her.

Often, during meetings about Apollo 8, George Mueller had pushed a piece of paper in front of Kraft and other senior NASA managers and asked them to estimate the probability of success at each phase of the flight; doing that would then yield the chances of success for the total mission. Kraft had been amazed to find his estimate to be within 1 percent of George Low's, and it's the one he gave Susan as she looked him in the eye with her question.

"How's fifty-fifty?" he said.

To Kraft, that seemed a hopeful number, given that the crew would be accomplishing so many new things at once: the first to fly the Saturn V, the first to journey to the Moon, and the first to confront the other myriad new challenges involved.

In fact, Kraft had misunderstood Susan's question. He thought she was asking about the odds of a successful mission, in which its objectives were met and the spacecraft and systems performed as designed. He and Low had figured those odds to be about 56 percent. If he'd understood that Susan was asking about the crew's chances of surviving the mission, he would have placed the odds higher.

And yet Susan, who believed her husband had no chance of coming home alive, was happy with a fifty-fifty shot that Frank would live.

"Good," she told Kraft, "that suits me fine."

That night, when Susan talked to Frank, she told him the cocktail party had been lovely.

In the first days of December, a new issue of *Time* magazine hit the newsstands. The cover image, set against a brilliant blue sky, showed two space travelers, one American, the other Soviet, sprinting toward the cratered lunar surface. Four words appeared on the cover: RACE FOR THE MOON.

The story inside summarized NASA's plans to send Apollo 8 to the Moon and the Soviets' push to send Zond 7 before the Americans could launch. Even at this late date, with just days remaining until the Soviet launch window opened on December 8, the race was too close to call.

In Moscow, the Soviets appeared to be celebrating early. Already, they had named a seventy-mile-wide crater on the far side of the Moon, photographed by Zond 6, in honor of two Soviet scientist brothers. In Florida, Anders was doing some naming of his own. Working from photos taken by unmanned spacecraft, he began assigning names to several of the most interesting and prominent craters never before seen by human eyes, ones he expected to see during his flight. Whether the International Astronomical Union would accept those designations once the crew had actually seen the craters remained to be determined.

It was around this time that Mueller asked the Apollo 8 astronauts to sign a statement confirming that they'd been properly trained by NASA. To Anders, that came as a bitter disappointment. Mueller was the boss—he should have been the one to tell the crew they'd been properly trained. Instead, he seemed to want a waiver in case anything went wrong.

By early December, the eyes of the world were trained on Kazakhstan. Cosmonauts were already at the launchpad there, awaiting the mission's final go-ahead.

On December 8, many at NASA held their breath. If the Soviets were going to send a manned spacecraft to the Moon, this was the forty-eight-hour window during which they would do it. More than a decade in the making, the Space Race was coming down to a matter of hours.

The first day passed.

The Soviets now had twenty-four hours to make their move to the Moon.

The second day passed.

The Soviets now had just a few hours remaining. If they were going to beat the United States they had to do it now.

By midnight, it was clear that nothing had happened at Baikonur. Technically, it was still possible for the Soviets to go, even as late as December 10. But by many assessments, if they were going to go in December, they would have gone already.

For the first time since the Space Race began, nothing stood between America and the Moon.

Nearly two hundred members of the media were accredited for an early December press conference in Houston with Borman, Lovell, and Anders. They were shown film of the astronauts training and were then allowed to ask questions of the three astronauts.

A reporter asked about recent comments by Sir Bernard Lovell in which the famed British astronomer had criticized NASA for taking undue risk by flying Apollo 8.

"I have the highest regard for him and I hope he has his telescope—his radio dishes—beamed on us," Borman said. "He's done a great job of tracking in the past."

When asked about the risks of the flight, Borman answered straight, as always.

"I think there are sensible risks. . . . If we really believe what we're doing is worthwhile, then we accept the risk. When we get to the point where we don't believe it's worthwhile, I'll quit."

In fact, Borman had already quit.

Several days earlier, he'd told NASA that he wouldn't fly in space after Apollo 8. Since becoming an astronaut in 1962, his mission had been simple: to beat the Soviets to the Moon. If all went well, Apollo 8 would do that. To risk another lunar journey just to pick up rocks or add a fraction to mankind's knowledge about the Moon didn't seem worth it after the battle had been won.

As he answered questions and made the media laugh, few

could have predicted this sudden turn in Borman's career. He was the consummate astronaut, a man for whom the mission always came first. Those who knew him best, however, might have guessed that there was an additional reason for his decision to hang it up after Apollo 8. By now, consciously or otherwise, Borman had come to see the stress his career placed on Susan, and he couldn't have any more of that.

After the press conference, the astronauts prepared to fly from Houston to Florida, where they would live for the remaining two weeks before their launch. This would be their last chance to say goodbye in person to their wives and children.

Borman and Lovell said farewell to their families at home, wished them a merry Christmas in advance, and told them they'd celebrate the holiday after they returned to Earth. Anders did the same, but then he gave Valerie a small package. It contained an audiotape. He asked that she play it in the event he didn't make it back. Anders was a private person and didn't tell anyone what he'd said on the tape. It began, "You children and your mother are the most important . . ." Much of the rest of it came down to this: expressions of love for Valerie and the kids; a reminder that he missed them already; a hope that Valerie would marry again in the future; and an assurance that he'd died doing what he wanted to be doing.

At the Cape, the men checked in to their new quarters, each getting a tiny room with little more than a steel bed and a steel desk, but with a large adjoining living room to share. Framed copies of classic paintings competed with lunar maps for space on the walls. It was a comfortable, if cramped, existence, and one deemed necessary by NASA to prevent the crew from catching bugs or viruses from the outside world that might short-circuit their ability to fly. The sole luxury came in the form of a personal chef.

No sooner had the astronauts arrived at the Cape than they

had to pack their bags again. Lyndon Johnson had invited them and their wives to the White House for a formal dinner and send-off, just twelve days before the flight. System checklists and countdown procedures swimming in their heads, the astronauts boarded a charter flight to Washington. Doctors didn't like the idea. The Hong Kong flu pandemic—which would kill more than thirty-three thousand in the United States alone in a six-month span—was reaching its peak, and the astronauts were supposed to be in quarantine. When a NASA doctor tried to object, LBJ issued a Texas-sized *Who the hell does he think he is?* For the crew of Apollo 8, there was a silver lining—a last, unexpected chance to kiss their wives goodbye.

As the astronauts flew to the White House, the family of one of the thirty thousand Americans killed so far during the fighting in Vietnam prepared for their own visit to the nation's capital.

On October 31, 1967, just a month before he was to return home, Captain Riley L. Pitts of the U.S. Army led his company on an assault of a Vietcong position in the dense jungle of Ap Dong, South Vietnam. After enduring withering fire, Pitts threw his body on top of an enemy hand grenade and waited to die. When the grenade failed to explode, Pitts moved his company forward, putting himself in the direct line of enemy fire until he was cut down in a hailstorm of bullets.

As the crew of Apollo 8 arrived with their wives at the White House, Capt. Pitts's widow, Eula, laid out a dark suit and a bow tie for her five-year-old son, Mark, and a fine white blouse for her seven-year-old daughter, Stacie, at their home in Oklahoma City. The next day, the president would make her husband the first African American officer ever to receive the nation's highest military decoration, the Congressional Medal of Honor. Millions of Americans considered astronauts to be the epitome of American courage. To Borman, Lovell, and Anders, that label better belonged to men like Pitts.

Joining the crew of Apollo 8 and their wives at the black tie gala were twenty other astronauts, Chris Kraft and Wernher von Braun, and former NASA chief James Webb, who was to be awarded the Presidential Medal of Freedom later in the evening. Also present was Charles Lindbergh, who'd stunned the world when he flew nonstop from New York to Paris in 1927. To many at NASA, despite his controversial political views, Lindbergh was a pinnacle aviation hero, a man who had taken to the skies to do the impossible.

Before dinner, a small concert was staged in the East Room. When Valerie Anders took her place in the audience, she was dismayed to hear dozens of people coughing and sneezing. *This is so stupid,* she thought. *They are putting this crew at risk.* And yet there was no escape for any of them. So they stayed, droplets of the Hong Kong flu and who knows what else atomizing into the room.

During dinner, Kraft got to talking to Lindbergh about airplanes, a nuts-and-bolts conversation between one of the great original aviators and an old flight test engineer. Seated nearby with the president, Borman stole glances toward Kraft's table, envying the conversation he was missing. He also noticed that LBJ seemed irascible, describing his annoyance at a press corps—and maybe an entire swath of the American public—whose criticism of his Vietnam policies seemed to have beaten him down. Listening to the president rail against the media, Borman felt empathy for Johnson, not just for the stigma of Vietnam that would attach to his legacy, but for how it must feel to be a man in the final days of his standing, knowing that soon he would never again matter in the way he once had.

Chapter Eleven

●

MY GOD, WE ARE REALLY DOING THIS

THE CREW OF APOLLO 8 RETURNED TO THEIR QUARTERS AT CAPE
Kennedy on December 10. Just eleven days remained until their
mission. Their schedule would be simple from this point forward: train in the simulator, study the flight plan, jog. At night,
when he could find a spare moment, Borman walked outside his
tiny bedroom and looked up at the Moon.

On December 15, at 7 P.M. EST, NASA began its official
launch countdown, five and a half days before the planned lift-
off. That gave everyone associated with the flight time to coordinate, and to fix any problems that might arise along the way.
As the clock started ticking that Sunday evening, Lovell bor-
rowed a car and drove sixty miles north along the Florida coast
to a town called Edgewater, where he rang the doorbell at a
house near the beach.

His mother, Blanch, opened the door. She lived here now.

Her seventy-third birthday was approaching, and her son had come to celebrate early. Over dinner, Jim explained the mission to his mother. Sitting shoulder to shoulder on the living room couch, Jim sketching out his rocket's trajectory, Blanch wearing her glasses and leaning in for a closer look, the two might have been in Milwaukee thirty years earlier, a mother and her young son who had only each other to rely on, each present for the other one's dreams.

The next day, NASA chief Tom Paine flew to the Cape to visit with the astronauts and to deliver an important message. After dinner—and a few drinks to loosen things up—Paine spoke frankly to Borman, Lovell, and Anders, with one final statement he wanted them to remember. He laid it out like this:

First, if any of them had any reservations going into the flight, anything they hadn't felt comfortable discussing with Chris Kraft or Deke Slayton or anyone else at NASA, even if it was nothing more than a feeling or an intuition that something wasn't right, he should feel free to bring those concerns to Paine, and he would personally see to it that the issue was addressed and fixed, no matter what, and without consequence to them.

Second, if the crew had any doubts or worries during the mission—with how the flight was progressing, with the function of the spacecraft or systems, with anything—he should feel free to abort the mission and bring the ship back early, and Paine would guarantee a seat on a subsequent flight as soon as possible. No one, he told the astronauts, would lose his chance to go to the Moon for ending a flight in the name of safety.

Borman, Lovell, and Anders thanked Paine for the offer, but none of them expressed any concerns about Apollo 8. All of them expected it would require a hell of a lot more than a feeling or an intuition to U-turn a spacecraft bound for the Moon.

On December 17, four days before scheduled lift-off, Marilyn Lovell and her four children landed in Florida and checked in to a beachside cottage. Valerie Anders, too, had managed to make the trip, catching a ride with a NASA contractor. Valerie had come only for the day, to squeeze in a final goodbye with her husband before returning home to be with their children.

The next day, Jerry Lederer, director of NASA's Office of Manned Space Flight Safety, spoke to a group of aviation enthusiasts in New York. Apollo 8, he said, had one safety advantage over the voyage undertaken by Christopher Columbus in 1492: "Columbus did not know where he was going, how far it was, nor where he had been after his return. With Apollo, there is no such lack of information." There was, however, the matter of complexity. "Apollo 8 has 5,600,000 parts and 1,500,000 systems, subsystems and assemblies," Lederer noted. "Even if all functioned with 99.9 percent reliability, we could expect 5,600 defects." For that reason, Lederer concluded, Apollo 8's mission would involve "risks of great magnitude and probably risks that have not been foreseen."

As darkness fell in Florida that night, Lovell took Marilyn out on a date. They didn't go to a restaurant or a movie, but rather to a place virtually no one on Earth could access, to see one of the newest wonders of the world. And no matter how high Marilyn looked when they reached the Cape, she still could not see where the great Saturn V ended, it just kept stretching upward, more than 250 feet taller than the rocket that had carried Lovell on his Gemini missions, a colossus lit white by floodlights against an inky black sky.

"I don't want you to worry," Lovell said, holding Marilyn's hand. "When we lift off, the rocket is going to tilt, it might even look like it's going to fall over, but that's normal, it's exactly how they designed it. Also, the Earth is going to shake in a different kind of way. That's normal, too."

By the morning of December 19, just forty-eight hours before lift-off, journalists were swarming at the Houston homes of the astronauts. Valerie and Susan were gracious, smiling for everyone, their hair and makeup done, all of them expressing support and admiration for their husbands. Valerie always wore the same dress for appearances on television—yellow, with a close-fitting waist and knee-length skirt. Her mother noticed and asked her about it. Valerie had to confess: It was the only good dress she owned. Her husband was about to become one of the most famous men in the world, yet he still earned military pay, about $16,000 per year (plus another $16,000 from *Life* magazine), which went only so far with five children to feed.

That night, Valerie decided to slip out of the house with three-year-old Eric to go for some groceries. She stole out the back gate and headed for the garage but was greeted in her driveway by an ocean of reporters and bursting flashbulbs. The next day, photographs ran across the country showing Eric in his mother's arms, sucking his thumb, along with the caption THUMBS UP FOR DAD! Valerie loved the photo, but she knew it meant she would be a captive in her own home from that moment forward.

On December 20, the day before the flight, the Soviets let the world know what they thought about Apollo 8. "It is not important to mankind who will reach the Moon first and when he will reach it," said cosmonaut Gherman Titov, the second man ever to orbit Earth. Not many in the Soviet Union were worried. Even with the American countdown clock at T minus 24 hours, few Soviets believed NASA would be crazy enough to launch.

That afternoon, Charles Lindbergh and his wife, Anne Morrow, joined the astronauts in Florida for lunch. Anders suspected the visit to be a public relations stunt arranged by NASA, but he changed his mind after hearing the passion in Lindbergh's questions about Apollo 8. After a few minutes, the four men—and Anne, also a pilot—were immersed in conversation about flying. Not one of the astronauts resented the imposition on his time. By now, there wasn't any sense in cramming more pages from a flight manual or checklist; with twenty hours to go until launch, you either knew your stuff or you didn't.

The conversation turned to spacewalking, and how it compared to the old barnstorming stunt of wing-walking. The Lindberghs were interested to hear that one's sensation of altitude decreased as one flew higher, until it hardly seemed to register in space (where the familiar scenery that helped people judge distance from the ground all but disappeared), and to learn that in space there was no up or down. The astronauts were equally interested to learn of a conversation Lindbergh once had in the 1930s with Robert Goddard, the father of modern rocket engineering. It was theoretically possible, Goddard had told Lindbergh, to design a rocket powerful enough to reach the Moon, but the money required to build it—as much as a million dollars—would likely keep such a wonder in the realm of science fiction. The astronauts had a good laugh at that one.

Lindbergh performed a back-of-napkin calculation after learning how much fuel the Saturn V required to send Apollo 8 to the Moon. "In the first second of your flight," Lindbergh said, "you'll burn more than ten times as much as I did flying the *Spirit of St. Louis* all the way from New York to Paris."

Later that day, Anders's childhood priest arrived at crew quarters. Father Dennis Barry had come to give Anders—a devout Catholic since childhood—communion. This visit annoyed Borman, who was growing edgier as the hours to launch counted down. The longer Father Barry stayed, the more irritated Borman grew. Finally, Borman snapped.

"Are you gonna take communion every thirty seconds before the flight?" Borman asked.

"No, Frank. He's just visiting," Anders said.

"Well, then, get rid of the guy!"

Borman was sorry he said it, even sorrier when he saw that his remark had hurt Anders. But he thought the crew didn't need distractions so close to launch.

Anders had more visitors coming. One was his thesis adviser and head of the Department of Physics at the Air Force Institute of Technology; the other was the man's brother, who was a Jesuit priest. He considered both to be very good friends. Around sunset, Anders took them to the parking lot outside crew quarters, where they all lay back on the hood of a car. By now, the sky had darkened, and the men picked out the slivered crescent of the Moon in the sky.

Early that evening, his last night on Earth before launch, Lovell sneaked away from crew quarters to visit Marilyn. She'd been at a party, but when Jim called, she hurried away and met him for a rendezvous at the cottage where she and her children were staying. There he kissed his kids and pulled out a photograph. Taken by one of NASA's unmanned lunar probes, it showed an angular mountain on the Moon. It was near the Sea of Tranquillity, one of the potential sites Apollo 8 would scout for a future landing mission.

"I'm going to name it Mount Marilyn," he said.

Wake-up would be at 2:30 A.M. The astronauts were to eat dinner, then go to sleep. Launch, at 7:21 A.M., was just twelve hours away.

After the meal, the men called their families to say good night and goodbye. Borman spoke first to his boys, then to Susan.

"Everything's going to be all right," he told her. "I'll be perfectly safe."

"I know," Susan said.

Before retiring, Borman knelt by his bedside to pray—the Lord's Prayer, then a request for a successful mission, and finally that he, Lovell, and Anders do their jobs well.

But he couldn't sleep, not for hours. He and his crew had been given only four months to train for the flight, and he was concerned about everyone's ability to perform flawlessly. More than anything, he dreaded the possibility of having to fly the backup mission—ten endless days in Earth orbit—if some anomaly was found after launch that spooked NASA into canceling the lunar part of the journey. He did not want to leave Susan a widow and his boys fatherless, but on this account, he didn't worry too much; he believed in the rocket and spacecraft the crew would be flying, and especially in the people who'd built and designed them. As Borman saw it, he would be flying with thousands of the world's best minds aboard.

Around midnight, ground crews began to pump each of the three stages of the Saturn V with liquid oxygen, an oxidizer necessary for combustion. Ribbons of white vapor danced as the warm Florida air boiled away drops of the liquid oxygen from vents in the tanks. Fuel was next. Three weeks earlier, the Saturn V's first stage had been filled with 209,000 gallons of highly refined kerosene. Now liquid hydrogen was added to the second-stage (260,000 gallons) and to the third-stage (69,500 gallons). Added to the 437,000 gallons of liquid oxygen in the stages, the fully fueled rocket would hold nearly a million gallons of propellant and would weigh 6.2 million pounds, all with the explosive potential of a small nuclear bomb.

At 2:36 A.M. on Saturday December 21, 1968, Deke Slayton knocked on the bedroom doors of the astronauts and told them it was time.

The men took hot showers, then walked down the hall for a cursory physical exam. After being pronounced fit by a team of doctors, they made their way to the breakfast room, where they were joined by Slayton, George Low, Alan Shepard, astronaut Harrison Schmitt, and two of their three backup crew, Neil Armstrong and Buzz Aldrin (the third member, Fred Haise, was working inside the spacecraft setting dials and switches).

In heaping quantities, the astronauts' personal chef served filet mignon and scrambled eggs (steak and eggs was the traditional send-off meal for astronauts), toast, coffee, and tea—a deliberately low-residue meal, and the last real, hot food that the astronauts would consume for the next six days.

After breakfast, the crew of Apollo 8 made their way to the

suiting room. If astronauts could have flown in thirteenth-century chain mail, they might have preferred it to NASA's space suit. All of them, however, understood its necessity. The custom-tailored one-piece suits could be pressurized, and they were made fireproof by Teflon cloth. Layers of Mylar, Dacron, and Kapton protected from heat, while other layers provided cooling and controlled pressure. In all, these suits contained more than twenty layers of protective materials, enough to do battle with a universe that becomes hostile to humans just a few miles above Earth's surface. Fully dressed, the suit's wearer looked like a futuristic version of the Michelin Man and walked a bit like Frankenstein's monster, but the suit could be his lifeline during a space flight if the cabin lost pressure.

A team of technicians, dressed in surgical masks to avoid spreading germs, descended on the astronauts, ordering them into their long johns and biometric sensors (to transmit physiological data to Houston), then helping them don their suits. Borman's equipment specialist would be Joe Schmidt, an all-around good guy, and the same sergeant who'd helped him into his pressure suit so many times at Edwards Air Force Base, where Borman had been a test pilot. The two were old hands at this kind of dance.

To enter the space suit, Borman had to shimmy and shake his way in through a tight zipper opening in the back of the garment, favoring no limb over any other lest the rest of him be left behind. After he popped his head through the neck ring, oxygen and cooling hoses were attached to blue (input) and red (output) valves at his torso. The next piece went on easily—a soft cap like the ones worn in the 1930s by barnstorming pilots, the men who gave rides to kids like Borman, Lovell, and Anders. (NASA's caps, however, were woven with state-of-the-art communications gear—no yelling above the wind required.) Gloves were affixed and secured. Finally, a transparent bubble helmet was attached to the neck ring. (Borman's head was so large that his helmet cost an extra $45,000 to build.) Now fully kitted up, with pure oxygen flowing into their suits from portable ventilators, Borman, Lovell, and Anders were already separated from Earth, the only three men on the planet who needed the planet no more.

On the way out of the suiting room, Lovell's technician gave him a pocket handkerchief to dress up his space suit, while Schmidt presented Borman with a small paper Christmas tree. The gifts weren't intended to be brought on board, but it was the thought that counted.

Carrying their briefcase-sized oxygen ventilators, the crew of Apollo 8 lumbered down a long hallway, waving to a photographer as they made their way to the elevator that would carry them to the building's exit. Television cameras and a small group of well-wishers greeted them as they left the Manned Spacecraft Operations Building and boarded the astronaut transport van for the first leg of their journey, an eight-mile ride to the launchpad. Emblazoned on the van's inside door was the figure eight insignia Lovell had designed for this mission, along with a reminder in bold red letters: NO SMOKING.

As the van made its way to the pad around 4:30 A.M., a NASA official drove Marilyn Lovell and her four children to a sand dune about three miles from the launch site, the closest NASA allowed observers to be for lift-off. The family were alone when they arrived, except for the coffee and doughnuts NASA had waiting for them. It was still dark. Marilyn pulled a pack of Pall Malls from her purse. Often, she and Susan Borman met in the mornings for a cup of coffee and a smoke, but Marilyn was solo today, so she shook a cigarette loose and lit up by herself. In Houston, the Borman and Anders families began to wake and get ready for a day glued to their televisions.

Upon reaching Pad 39A, the astronauts exited the van and bent backward, straining in their suits and helmets to get a view of the behemoth before them. From a distance of just a few yards, the Saturn V was mythically tall. The idea that it could move, never mind fly, seemed impossible from up close.

The astronauts walked to the middle of the launchpad, then

took two elevator rides, the first to the zero level of the service tower, the second up the crisscrossed steel beams of that tower thirty-two stories into the air. Then they walked across an access arm to a small loading area, where technicians would make a final check of the space suits. From there it was a short walk for the astronauts across a small metal bridge and into the space-craft. It was 4:58 A.M., still dark outside. From his vantage point, Lovell couldn't help but think of the old astronaut joke—*How does it feel to sit atop a vehicle built by the lowest bidder?*

A NASA staffer gave the signal for the astronauts to start load-ing. Borman went first and, after some maneuvering, settled into the left-hand seat of the command module, lying flat on his back as an airline pilot would if his airplane were tipped back onto its tail. A technician gave Anders a hug, then sent him, too, into the spacecraft, where he took the right-hand seat in the small cabin. As Anders worked to get himself settled, Lovell looked down to the ground several hundred feet below. He could see the lights of the press corps as they arrived at their designated sites, and all of a sudden it hit him: *These NASA people are serious. They're going to send us to the Moon. My God, we really are doing this.* He took a deep breath, then walked across the bridge, put his feet through the hatch of the spacecraft, and lowered himself into the seat between Borman and Anders.

Technicians closed and secured the hatch on the Apollo 8 spacecraft at 5:34 A.M. Inside the cabin, the countdown clock read T minus 2 hours, 17 minutes and counting. Lying flat on their backs, there wasn't much the crew could do to help things along. Borman wished NASA could just get the damned thing into the air, but knowing that wasn't possible, he wished for something more realistic—that the launch would actually occur. He didn't want another episode such as John Glenn endured in the Mercury days, when his flight was scrapped with twenty-nine minutes to go. If a guy was going to suit up and climb aboard with his nerves on edge, the least a rocket could do was go up.

In the command centers at the Cape and in Houston, controllers readied for launch, settling in to legacy aromas of stale pizza and burnt coffee, checking their consoles and lists and running through their responsibilities as they had done for the past four months in their offices, at dinner with their families, in bed after their wives had fallen asleep. The Cape would be in charge of the launch (since they would be on scene), then turn over command to Houston shortly after lift-off. Both command centers vibrated as hundreds of controllers moved into position.

At the helm would be Flight Director Cliff Charlesworth, in charge of the flight from the ground. Charlesworth could take any action he deemed necessary to ensure the safety of the crew and the success of the mission. The CapCom—always a fellow astronaut—would do most of the communicating with the crew of the spacecraft as it flew, and he would be the crew's advocate in Mission Control. The facility would operate around the clock in eight-hour shifts, as long as the mission lasted. Each team of flight controllers was designated by a color. The primary Cap-Com, Mike Collins, was on Charlesworth's Green team and would cover launch through to the historic TLI maneuver.

Overseeing them all in Houston would be Chris Kraft, the director of flight operations. Even now, Kraft might have been the most nervous of them all. He'd been with NASA since its inception in 1958. More than anyone else, he knew how much could go right—and wrong—when men left Earth.

The astronauts occupied themselves by checking switches, confirming checklists, and eavesdropping through their headsets on launch personnel. They could not hear NASA public affairs officer Jack King, whose baritone voice and slight Boston accent had kept the world updated live on launch countdowns since the Mercury flights. Borman, Lovell, and Anders shivered in their space suits, their cabin freezing in the still-chilly morning air.

The astronauts could do little more than wait. Through a tiny porthole in front of him, Borman watched two seagulls flying around the spacecraft and checking out the strange, tall bird—

the Saturn V—that now shared their sky. From his middle seat, Lovell scanned the instrument panel and admired the detail of the Apollo simulators; nothing inside the command module looked or felt different from what the crew had practiced with on the ground. And in a testament to the cool that runs through the bloodstream of fighter pilots, Anders fell asleep, ready to awaken when things got good.

By 7 A.M., network coverage of the launch had gone live on televisions and radios across America and the world. As the countdown clock ticked under an hour, a crowd gathered around the color television in the living room of the Borman home. Susan, her two sons, Frank's parents (who had arrived at three A.M. and now fidgeted nervously on the couch), and family friends had come to watch the launch. Joining them were a *Life* magazine photographer, along with the wives of seven other astronauts, some of whom had brought deviled eggs and champagne. Though she smiled for newspaper photographers while scratching the tummy of the family's shaggy dog, Teddy, Susan's insides were in knots. At every chance, she hurried back to the squawk boxes NASA had installed in her home so that she could listen directly to the communications between Apollo 8 and Mission Control.

At the Anders home, Valerie scrambled to round up her five children, seating them atop the toy box in the playroom where the family kept their new color TV. Joining them were Bill's aunt and uncle, several family friends, and the wives of some of the other astronauts (Bill's parents were at home in San Diego to watch the launch). Valerie tried to stay in the moment, absorbing everything, even the fear, full of hope.

Among those watching the countdown from behind a giant window at the Launch Control Center at the Cape was backup crew member Neil Armstrong, who couldn't get over the moxie NASA had shown in conceiving the mission. The Saturn V had never been flown with men aboard and had suffered profound problems on its second and most recent test. To put a crew on that rocket now, and to point that crew at the Moon, seemed astonishingly aggressive—and wonderful—to him.

Just twenty minutes remained until launch. For miles along the Cape, thousands of cars and motorcycles and buses and campers jammed the beaches and roadways, a quarter of a million people standing on hoods or in sand or on one another's shoulders, craning their necks for a view of the rocket, passing binoculars back and forth, checking their watches every few seconds. An eighty-year-old woman from South Dakota, who'd traveled in her son's trailer to witness the launch, said, "Those men, what they will do! And I have lived to see it. I am still alive to see it."

"T minus 7 minutes, 30 seconds and counting, still aiming toward our planned lift-off time," King told a riveted nation.

In the morning light, the view from the spacecraft became clearer. For several minutes, Anders watched a mud dauber wasp build a nest on the capsule window.

At T minus 5 minutes, the access arm and loading area pulled away from the spacecraft and retracted. At Mission Control in Houston, Chris Kraft stared at a giant color screen that was broadcasting a view of the spacecraft. He had always considered launch to be the riskiest part of manned spaceflight, and that was true even with proven rockets. Now, feeling scared to death, he watched as his agency prepared to catapult three good men from the planet aboard the most powerful machine ever built, despite the fact that this machine had never lifted a living thing, not even a mealworm, off the ground.

At T minus 3 minutes, 6 seconds, computers took over full checkout of the rocket.

In the Borman and Anders homes, hands were squeezed tight. On the sand dune, Marilyn Lovell huddled with her children. Photographers from various media outlets were stationed with all three families, pressing shutter buttons and swapping film rolls as fast as they could, desperate not to miss a moment.

In Mission Control, Flight Director Cliff Charlesworth and CapCom Mike Collins watched the monitors. George Low, the man who'd conceived a mission the Soviets still didn't believe would fly, breathed deeply as the clock showed just one minute remaining.

At T minus 60 seconds, all three stages of the Saturn V were fully pressurized.

"Twenty seconds," Jack King announced to the world. "We are still Go at this time."

Storms of white vapor began to billow near the base of the rocket, liquid oxygen boiling off during the Saturn V's final moments on Earth. Inside the spacecraft, Borman felt the rocket sway a bit in the wind.

"T minus fifteen," King called, "fourteen . . . thirteen . . . twelve . . . eleven . . . ten . . ."

Heart pounding, Borman's left hand remained gripping one of the spacecraft's controls, ready to twist it to the left and abort the mission in case of a catastrophic problem. The three men listened to propellant pumping through the engine manifolds. On the beach, Marilyn reminded herself that the rocket would lean when it took off, just as Jim had warned.

"Nine . . . We have ignition sequence start, the engines are armed!" King said, as a fury of orange-yellow flames lit beneath the rocket and exploded against the launchpad.

"Four . . ."

"Three . . ."

Flames spread from beneath the rocket and erupted out to the sides, a typhoon of fire awakened and screaming as the ground began to shake.

"Two . . ."

A man-made thunder crashed into people and windows and buildings for miles around.

"One . . ."

Borman loosened his grip on the abort handle. He would

have rather died than twist it by accident in the violence unfolding beneath him.

"Zero. We have commit . . . We have . . ."

King paused for a moment, as if he didn't quite believe what he was seeing.

Susan Borman, Marilyn Lovell, and Valerie Anders didn't breathe.

At 7:51 A.M., King called it.

"We have lift-off."

And Apollo 8 began to move.

Chapter Twelve

●

LEAVING HOME

APOLLO 8 STRAINED TO SEPARATE FROM EARTH, EXPLOSIONS OF smoke and fire channeled to the sides of the launchpad by a massive, wedge-shaped flame deflector designed to prevent the fire from rebounding back up into the rocket. Four hold-down arms held the rocket in place, waiting for the Saturn V's engines to attain perfect, proper thrust. A fraction of a second later, these arms released to allow the machine to ascend. Five swing arms, each weighing more than twenty tons, remained for a split second longer with their connections to the rocket intact. As the Saturn V began to rise, they, too, withdrew, and the six-and-a-half-million-pound beast broke free from its bonds.

A half inch off the pad, it was already too late for the rocket to settle back safely if something went wrong. System design would not allow shutdown of the engines for thirty seconds even in the event of catastrophic failure, since this would cause the Saturn V to fall back onto the launchpad and explode. If there was any major failure now, Borman would have to twist the abort handle and allow the rocket-propelled escape tower at the top of the spacecraft to pull the command module free and hurtle them out to sea.

There was no failure. Each engine was erupting and functioning just as von Braun had envisioned, producing a combined 7.6 million pounds of thrust, or 160 million horsepower—enough energy to power the entire United Kingdom at peak usage time—as the rocket began to inch upward. Blocks of ice formed by the supercool liquid oxygen in the rocket's first stage shook free from the Saturn's torso and splintered into a white confetti that rained into the firestorm below.

A few feet off the pad, the Saturn V began to lean away from the support tower. This was the maneuver Lovell had described to Marilyn, designed to keep the vehicle safe from wind gusts that might throw it back into the tower. Down the nearby beaches, the ground began to shake, and people's chests were pounded by the pressure waves, and it spread out at the speed of sound for miles around the Cape.

Inside the command module, the noise had already become deafening for the astronauts, their headsets rendered useless for communicating with the Cape or with one another. Borman and Lovell could sense the slowness of the acceleration due to the sheer weight of the Saturn V, a much different kind of movement than they'd experienced from the nimble Titan II rocket that had powered the Gemini program just two years earlier. But it wasn't just the speed that was different. The cabin shook so violently that Anders believed the rocket's fins were grinding through the girders of the launch tower and being shorn off. He tried to find an instrument or a gauge to monitor, something that would confirm the disaster unfolding beneath him, but his head was being shaken with such force he couldn't focus or even think, and even if he could have, he never could have communicated any information to Borman, either by speaking or signaling, since he was no longer in control of his body and his arms had turned to lead. None of this had been predicted or simulated. In the mountains of books and reams of papers, no one had mentioned that even before the rocket cleared the tower, the world inside it would be coming apart. *Holy shit,* Anders managed to think as the bodies of the three astronauts were ragdolled against their straps, *what the hell is going on?*

And the rocket still hadn't cleared the launch tower.

Groaning under its own weight, the Saturn V began to move higher, bending farther away from the tower as the spitting tail of flame grew longer. The flight was now just ten seconds old, but Anders already felt like a rat in the jaws of a giant, angry terrier, helpless to do anything but hang on and breathe while the five massive F-1 engines constantly swiveled their thrust to keep the 363-foot rocket from toppling over. Again, Anders tried to pick out instruments to get an idea of what was happening, but the rocket kept thrashing him into Lovell, against the wall, into his straps. The crew had trained for hundreds of hours for every kind of emergency, but NASA's simulators were not the kind of dynamic, multiaxis machines that could come close to approximating such violence. If an engine had fallen off or exploded, if the rocket had been engulfed in flames, if any number of disasters had been unfolding, the crew wouldn't have known about it, and they wouldn't have been able to hear Mission Control tell them about it, either.

Still, Borman kept his hand clear of the abort handle.

To Anders, it seemed that the flight had already lasted an hour when he and his crewmates heard the first, faint transmission in their headphones, a call from the launch operations manager at the Cape that conveyed a simple but essential piece of information.

"Tower clear."

The call had come thirteen seconds into the flight. At home, Borman's seventeen-year-old son, Fred, watched on TV. He'd never known anyone as committed to his work as his father, a man he still saw as a fighter pilot at Edwards Air Force Base, a man who refused to crash in machines that crashed all the time. So Fred was calm today as the rocket climbed, just as he had been during Gemini 7, just as he had been every time there had been sirens and black smoke in the sky at Edwards. *Just keep going, Dad,* Fred thought. *If you just keep doing that, everything's going to be fine.*

Borman radioed back to Houston, which had just assumed command from the Cape now that the tower was clear. His voice quaked along with the rocket: "Roll and pitch program."

He was confirming that the vehicle was turning to head out to sea exactly as required.

"Roger," answered CapCom Mike Collins.

The punishing cacophony began to diminish as a seagull—the same one Borman had seen before lift-off?—flew past the ripples of sound and smoke made by the rocket.

"How do you hear me, Houston?" Borman asked.

"Loud and clear," Collins answered.

Apollo 8 climbed higher, riding a column of fire into a brilliant blue sky. In Houston, controllers watched for any sign of catastrophe, ready to relay abort instructions to Borman, but all they saw were solid reports from their consoles. Inside the spacecraft, Anders could feel the ride smoothing out, and while he hoped the rocket's fins hadn't been ripped off by the tower, he figured he was probably okay given that, by all indications, he was still alive. *My God,* he thought. *If we missed that in our training, what else have we missed?*

Around the world, millions watched as the rocket pushed higher into the sky. In Houston, Susan Borman forgot for a moment that Frank was going to die on the flight. Instead she sat, still in her fancy cream-colored dress with a string of pearls around her neck, knees up against her chest, hands clasped, awestruck; to her, seeing Apollo 8 launch was like watching the Empire State Building leave Earth.

The Saturn's engines continued to burn 15 tons of propellant per second. Even as the ride smoothed, g-forces built inside the cabin, pinning the crew against their seats. Every movement and adjustment by the rocket was preprogrammed and executed by the Saturn's onboard computer, so the crew could do little more than study the readouts on their instrument panels, watching the rocket's velocity climb and the countdown grow closer to the end of the Saturn's first stage. On nearby beaches, people watching the rocket ascend stood on cars and cheered and waved American flags. One of them was Lovell's fifteen-year-old daughter, Barbara, who felt as if the Earth was cracking beneath her from the rocket's shock waves.

One minute into the flight, Apollo 8 reached the speed of sound—767 miles per hour—and an altitude of about 24,000 feet. At that point, the roar and crackle generated by the interaction between engine exhaust and air could not move fast enough to catch up to the spacecraft, and the cabin grew quiet, the hum of its instrument panel the only noise Borman could detect. To him, it now sounded as if he was flying an unpowered glider.

Outside, the Saturn V's first-stage engines still burned furiously as the rocket continued to gain speed as it rose an arced out to sea. In less than twenty seconds, the vehicle would reach a state known as max Q, the moment at which an airframe is subjected to maximum aerodynamic pressure. At Mission Control in Houston, Kraft felt his stomach twist. For months, he'd worried about the effects of max Q on Apollo 8.

It happened about one minute and nineteen seconds into the flight, at an altitude of 44,062 feet and a speed of about 1,500 miles per hour. The Saturn's five engines bellowed flames still visible to spectators up and down the Florida coast. In the press area, television cameras were now angled almost vertically on their tripods. If the ship was going to break apart, now was a likely time for it. But Apollo 8 passed this stress test and continued to soar into the sky.

The spacecraft now gained speed and altitude fast. Two minutes into the flight, it was traveling at 3,300 miles per hour and was 100,000 feet above Earth. Here, the atmosphere had grown so thin that the rocket faced little risk of disaster caused by aero-dynamic stress, but the next hurdle awaited. In less than a min-ute, the propellant for the first-stage engines would run out. At that point, the stage would have to be severed and allowed to fall away from the ship.

The engines continued to burn staggering amounts of pro-pellant, causing the ship to grow lighter and g-forces to increase, pressing the astronauts into their seats with up to four times the force of gravity on Earth, making each man's arm feel as if it weighed about thirty-six pounds. At an altitude of about 215,000 feet, the spacecraft reached a speed of 4,236 miles per hour. With the onboard clock nearing two and a half minutes' elapsed time, the first stage shut down, explosives fired, then retro rock-ets ignited, separating the first stage from the rest of the Sat-urn V and enveloping all of Apollo 8 in a cocoon of fire. To many of those watching from the ground and on television, it appeared that the entire ship had exploded, but it was just the precursor to the first stage falling back toward the Atlantic, glow-ing a brilliant goodbye.

Inside the spacecraft, the sudden shutdown of the first stage caused g-forces to drop from four to zero almost instantly. Hav-ing been severely compressed, the 363-foot tower of aluminum alloy suddenly sprang back, flinging the astronauts forward with explosive force. By instinct, Anders threw up his hand in front of his face to prevent being catapulted through the instrument panel, but by that time, the five J-2 engines of the second stage had kicked in and the acceleration threw Anders's outstretched hand back so hard against his head that the wrist ring on his glove carved a gouge in his helmet. As with the launch itself, simulations hadn't come close to preparing the astronauts for the violence of this moment.

Pinned back once again by the force of five screaming en-gines, the crew began to check instruments to make sure all was okay. Out of the corner of his eye, Anders glanced to check whether Borman or Lovell had noticed the gouge in his helmet, the sure mark of a rookie astronaut. Thankfully, it seemed they hadn't.

A few seconds after the first-stage booster fell away, Borman prepared to get rid of the other end of the vehicle, the thirty-four-foot-tall spire-shaped escape tower and conical boost protective cover, which had ridden atop Apollo 8, poised to rocket the command module away from the Saturn V in case of emergency. Cutting it loose meant a great saving in weight. If an abort was necessary after losing the escape tower, the crew would use propulsion systems built into the command and service modules to separate from the Saturn V, redirect their course, and ride the command module back to splashdown.

Borman threw a switch, causing a small rocket motor to jettison the escape tower. Instantly, the cabin was awash in sunlight, its five windows no longer obstructed by the boost protective cover. Now at an altitude of 300,000 feet, the astronauts could see the curvature of Earth against a blue sky that melded into the deep purple-black of onrushing outer space.

Two minutes later, Apollo 8 reached 100 miles altitude as it arced almost horizontally over Earth. The ship was now 350 miles downrange of Cape Kennedy and just about high enough for its planned Earth orbit. Speed, however, was another matter. To achieve orbit, the spacecraft needed to reach approximately 17,425 miles per hour; anything less and Earth's gravity would pull it back down. At the moment, six minutes into the flight, it was traveling only 10,000 miles per hour. Apollo 8 needed a big push, and that was the job of the five second-stage engines. Borman could see indicators of the ship's speed galloping forward on a five-digit readout on the instrument panel. If the Saturn's second stage failed now, the crew could use the rocket's single third-stage engine to get them to orbital speed—but if that happened, they wouldn't have enough propellant left to send the spacecraft on to the Moon, and Apollo 8 would become a days-long Earth-orbital checkout mission. That was the scenario Borman dreaded.

So far, however, the second-stage booster was flying true and smooth as it pushed the spacecraft's speed from 10,000 to 14,000 miles per hour in just two minutes' time. The five engines needed to burn for only another forty-five seconds before falling off and giving way to the third stage. Even for a conservative pilot like Borman, those forty-five seconds seemed a near certainty now.

And then he felt something go wrong.

The rocket beneath him started to shake furiously—a pogo— a problem similar to the one that had afflicted the unmanned Apollo 6 on the Saturn V's second and most recent test flight. Stresses created by pogo could damage or even tear apart the rocket. Von Braun and his engineers believed they'd worked out the issue, but this was exactly the kind of thing no one could know for certain without making another test flight, and there hadn't been another test flight after Apollo 6.

As a longtime fighter pilot and test pilot, Borman didn't spook easily. Now he was concerned. But there was nothing for the crew to do now except hope that the Saturn V could endure the pogo for another forty seconds until the second stage burned itself out and separated from the vehicle.

The ship continued to shake even as it gained speed by hundreds of miles per hour every few seconds. With just nine seconds to go before the engines of the second stage were to cut off, Borman radioed to Collins in Houston.

"The pogo's damping out."

Collins barely had time to respond before the engines shut down as scheduled and the Saturn's second stage separated from Apollo 8 and began its long fall back to Earth, where its remains would sink into the Atlantic, just as the first stage had. A fraction of a second later, the Saturn's third stage, the S-IVB, kicked in, creating a flash Borman could see through the hatch window. In just three seconds, it attained full thrust, but the push was almost gentle compared to those of the first two stages, as the rocket no longer needed to overcome air resistance and was using just a single J-2 engine (which itself was less prone to vibration). Despite the pogo episode, Borman remained impressed by the beauty of the rocket's engineering.

In a matter of moments, the spacecraft passed over Bermuda and reached 95 percent of the velocity needed for Earth orbit. If all went well, it would take just over a minute to add the last 5 percent needed to give Apollo 8 enough speed to achieve that almost mystical equilibrium between falling back to Earth and flying off into endless space.

By now, the stack that comprised Apollo 8 was barely one-third as long as it had been at launch, and millions of pounds lighter. Driving this relative featherweight, the single third-stage engine pushed the spacecraft even faster, to a speed of 17,425 miles per hour, before cutting out eleven minutes and twenty-five seconds into the flight.

Moments later, Apollo 8 was in orbit around Earth.

In Houston, Kraft, Collins, and rows of others exhaled. This was the first time the Saturn V had been asked to deliver men into space, and it had succeeded.

The Saturn's third stage had worked beautifully, but unlike the previous two stages, it did not separate from the spacecraft. It would be needed again, in about two hours and forty minutes—this time to send Borman, Lovell, and Anders to the Moon.

Apollo 8 settled into an easy orbit around Earth. It would circle the planet every 88 minutes and 10 seconds, with a perigee (closest point to Earth) of 113 miles and an apogee (farthest point from Earth) of 119 miles. With no more external forces pushing them forward, the astronauts experienced weightlessness inside the cabin. Borman and Lovell were old hands with the sensation, which was caused not by a lack of gravity but by being in the constant freefall that is orbit. Anders couldn't wait to give it a go.

But it wasn't time to unstrap yet. Apollo 8 was in a parking orbit around Earth for a reason—to allow the crew and controllers to confirm the proper function of the spacecraft's systems and hardware. Only if all looked flawless would Borman, Lovell, and Anders be allowed to leave the planet.

While the astronauts checked their spacecraft, their wives met the press back on Earth. Standing on her front lawn, still wearing pearls and with her hair in a perfect blond flip, Susan Borman told reporters, "This is very much different from Gemini 7. The magnitude of this entire thing is very difficult to comprehend and hasn't sunk in on me. I've always been known as a person who had something to say. Today, I am speechless. I'm too emotionally drained to talk."

Nearby, a tearful Valerie Anders, wearing a bright yellow raincoat and black patent leather shoes, told the press that no one had spoken in her home during lift-off. "It was beautiful and it has remained beautiful," she said.

When the interviews concluded, the women returned to their squawk boxes to continue following their husbands. After Susan closed her front door, she told a guest, "I don't know if I can sit here for the next two hours." She lit a cigarette. She'd made a solemn resolution to quit—as soon as Frank returned from the Moon.

The astronauts had just over two and a half hours to check out the spacecraft and its systems—less than two full revolutions around Earth—so they needed to make every minute count. Borman knew that Anders, a rookie, would be tempted to steal glances out the window and had taken to warning him during training not to try it during this critical time. When Borman glanced to his right, he saw Anders hard at work checking systems.

It took Anders a full three or four seconds after that to begin peeking out his window. A few moments later, Borman and Lovell were looking, too.

And there was Earth, a kaleidoscope of color turning in a black sky. Swirling cotton-white clouds revealed brilliant blue oceans beneath their breaks, while brown and green stretches of forest and jungle covered entire countries. Thin bands of blue followed the curvature of Earth, an incandescent skin of atmosphere come alive in a sea of darkness. Just a portion of the full sphere could be seen at this altitude of 116 miles, but already the astronauts could pick out the outlines of continents.

The three men checked and rechecked the spacecraft's switches, dials, buttons, readouts, valves, circuit breakers. They confirmed that the command module's maneuvering system was operational and healthy—they would need it for reentry to Earth's atmosphere on the return trip from the Moon. And they verified that the spacecraft's critical systems, including electrical and environmental (for oxygen, cooling of electronics, and temperature), were working perfectly.

Around seventeen minutes into the flight, Lovell was the first to unstrap and leave his seat. He headed to the Lower Equipment Bay to use the spacecraft's optics to make sightings of the stars. This would refine the guidance system's determination of direction. Throughout the mission, ground engineers would use the radio signal to the spacecraft to work out where Apollo 8 was and where it was going. If the radio equipment failed, Lovell would be the one in charge of getting Apollo 8 to the Moon—and back to Earth. He would spend a lot of time at this optics station practicing the use of the spacecraft's sextant and telescope to measure the angles between stars and the horizon of Earth or the Moon. For centuries, sailing ships had navigated by using sextants to make similar measurements. Even in the space age, it was hard to improve on these ancient techniques. This was Lovell's first chance to give the system a go.

He felt nauseated almost immediately. That hadn't been a problem during his Gemini flights, but the Apollo spacecraft was much larger and gave him more opportunity to turn his head and move around—a likely cause of his queasy feeling. He warned his crewmates—don't turn your heads when you first leave your seats—then steadied himself and allowed it to pass.

Near twenty-five minutes into the flight, after all was confirmed to be operating smoothly, the crew were allowed to remove their helmets and gloves. Doing so would help them move about the cabin and check the various switches and systems. A few minutes later, Anders got out of his straps and left his seat. For the first time, he was experiencing weightlessness.

"Hey, it's like sitting on an ice rink, isn't it?" he said to his crewmates.

Despite their liberation, the crew still needed to stay in their suits. Anders gestured to the others to hand him their helmets so he could stow them. He was surprised to see that each one had a gouge just like the one he'd made when trying to protect his face during the violent shutdown of the rocket's first stage. *I guess we're all rookies on a Saturn V,* he thought.

Forty-five minutes into the flight, Apollo 8 approached darkness for the first time as it moved eastward at more than 17,000 miles per hour, closing in on the side of the planet where it was currently nighttime. Outside the cabin windows, the crew could see lightning flashes from a storm down below, as if a thousand paparazzi had gathered to take pictures of them from the clouds.

They were still looking when Lovell called out into the cabin.

"Oh, shoot!"

"What was that?" Borman asked.

"My life jacket."

The astronauts all wore life vests over their suits, in case the spacecraft had to bail out over the Atlantic during launch. Lovell had caught his on something, and it had inflated with carbon dioxide. The vest had earned its nickname—Mae West—after the buxom Hollywood starlet. Even as Borman and Anders smiled, everyone knew they had a problem. Inflated, the Mae West took up space in a cabin where every square inch counted. Deflating it, however, would release levels of carbon dioxide that might be too much for the environmental control system to remove, and that might prove harmful to the crew.

Borman wasn't angry at Lovell, just irritated, and he'd have been irritated at himself if he'd done it. The challenge now was to figure out a solution. Anders hit on one a few seconds later—an answer of old-school simplicity. When the crew got a chance, they would deflate the vest through the urine dump, the same valve through which they got rid of liquid bodily waste. For now, they simply tied it down and got on with their work.

About ninety minutes into the flight, Apollo 8 passed over the Pacific Ocean toward Southern California. Out his window, Anders stole a glimpse of Los Angeles, picking out Santa Monica and Pacific Palisades, and then his hometown of San Diego, and the beaches where he first met Valerie. He felt a moment's envy— Borman and Lovell had enjoyed these spectacular rolling views for two weeks aboard Gemini 7. For Anders, the view would last for just another eighty minutes. Officially, he wasn't even supposed to be looking. Anders checked to make sure Borman wasn't watching, then stole a long glance out the window and let it settle into his memory.

It took Apollo 8 less than eleven minutes more to cross the United States. In Houston, Flight Director Cliff Charlesworth told Mission Control that the S-IVB, the Saturn's third-stage booster, looked good for translunar injection, the maneuver that would propel the spacecraft out of its orbit around Earth and on to the Moon.

To pull it off, Apollo 8 needed to accelerate from its current speed of about 17,400 miles per hour to nearly 24,250 miles per hour. That boost would be accomplished by the single J-2 engine on the Saturn's third stage, which would be reignited and burned for nearly six minutes. Doing this would not, as many believed, cause the spacecraft to leave Earth orbit; rather, it would simply change the shape of Apollo 8's orbit around Earth from a near circle into a highly elongated ellipse, one that would stretch all the way from Earth to the Moon and back.

The exact moment of the engine's firing, as well as its thrust, direction, and duration, depended on complex mathematics designed to put the spacecraft at just the right point where it could slingshot around the far side of the Moon and make a free return to Earth if necessary, all while accounting for the movements of Earth, the Moon, and the spacecraft itself. Row after row of controllers in Houston, in shirtsleeves and ties, seated at consoles crowded with monitors, buttons, levers, and dials, needed to analyze data pouring in from the spacecraft, trying to determine whether Apollo 8 looked ready to go to the Moon.

"How does it feel up there?" Collins radioed to the crew.

"Very good, very good," Borman replied. "Everything is going rather well. The Earth looks just about the same way it did three years ago."

Mission Control got a kick out of that one.

Meanwhile, in Hawaii, NASA had made it known that the translunar injection burn would occur almost directly over the islands. If the night stayed clear, locals might see Apollo 8's third-stage engine ignite as it hurled the spacecraft toward the Moon.

Just twenty-five minutes remained before the scheduled burn. Borman could picture Chris Kraft as he prepared for the historic maneuver, chewing on a stale cigar, watching a console over a controller's shoulder, processing the flood of information pouring in. Even now, no matter how far along things were, no matter how many things kept going right on board, Borman still couldn't quite accept that Kraft, or Flight Director Charlesworth, would really go through with this. As Borman began to stow equipment, Charlesworth put out his Lucky Strike cigarette and began to "call the roll," going console by console to ask his men at Mission Control for a "Go" or "No Go" verdict on translunar injection.

One by one, they gave him the same answer: "Go."

Charlesworth looked to CapCom Collins.

It was up to Collins to pass along Houston's decision to Apollo 8. For the first time, mankind was about to leave its home planet in search of a new world. To Collins, a man to whom history mattered, the event required words worthy of the moment, a statement that not only captured this cutting of the cosmic umbilical cord but would remind future generations that humankind understood the magnitude of what it was about to attempt.

He radioed the spacecraft.

"Apollo 8, Houston."

"Go ahead, Houston," Borman replied.

"Apollo 8. You are Go for TLI. Over."

And that was it.

Shit! Collins thought. *Here we are at this instant in history that shall forever be remembered, and it's just some guy saying "You are Go for TLI."*

And yet that was the way men who lived faster than sound communicated—just enough to give direction and then get out of the way.

"Roger," Borman replied. "We understand. We are Go for TLI."

Twenty minutes remained before the third-stage engine would reignite and begin the translunar injection burn. At home in Houston, Susan Borman stayed attached to her squawk box. She wished for privacy at this moment; instead, documentary filmmakers had their cameras trained on her, along with microphones placed around the house. Susan had protested the invasion, but Frank told her it was part of NASA's plan and couldn't be helped. Valerie Anders faced the same scrutiny before the cameras but seemed less put off, at least for now.

The astronauts spent the next several minutes making final checks, strapping back into their seats, and preparing for the translunar injection burn. (They wouldn't need their helmets, since they didn't expect a breach in the cabin and a loss of pressure.) About ninety seconds before the engine was to fire, a light came on in the command module indicating the final countdown to ignition. In the Soviet Union, many in the space program still could not believe what was occurring. *We could have done this first,* one cosmonaut thought. *Only the indecisiveness of our chief designer caused us to fall behind.*

Just sixty seconds remained, but the crew could still cancel the burn by throwing the Inhibit switch if things didn't look right. At eighteen seconds, they'd have no choice but to allow the engine to light.

Eighteen seconds now remained.

Kraft's heart pounded. He well remembered the second and most recent test flight of the Saturn V, when the third stage had flat-out failed to restart. If that happened again, Apollo 8 would be fated to an orbit around Earth—and the lunar mission would have failed.

Eight seconds before ignition. Liquid hydrogen began to run through opened valves and to the engine as Borman counted down.

". . . Four . . . Three . . . Two . . ."

Liquid hydrogen and liquid oxygen flooded into the engine's combustion chamber. An indicator in the cabin lighted up brightly, telling Borman and crew ignition was imminent.

In Hawaii, hundreds of people gazed upward. All they could see was a pinpoint of light.

"Ignition!" Lovell said.

The J-2 engine fired, pushing the astronauts gently back into their seats. In Hawaii, observers saw the tiny point of light explode into a giant streak of flame, a man-made comet glowing bright across the dark veil of sky. This was the acceleration that would be necessary to extend the spacecraft's orbit out to where the Moon would be in about sixty-six hours' time. At the moment, the Moon was farther back in its orbit around Earth and nowhere near where Apollo 8 was pointing. But if everything went right, the Moon, itself streaking through space at an average speed of 2,288 miles per hour, would arrive just in time to rendezvous with the little spacecraft.

While the astronauts monitored systems and prepared to fly

the vehicle manually in case the engine or steering system failed, Apollo 8 began to pitch up from its orientation parallel to Earth and climb higher away from its home. G-forces increased from 0.7 past 1.0 as Lovell watched the digits on the cabin's velocity indicator, which seemed to spin upward like the wheels of a slot machine.

Just over two minutes into the burn, Lovell reported a speed of nearly 20,000 miles per hour, faster than humans had ever moved. Over the next three minutes, the craft would need to exceed 24,000 miles per hour to achieve the desired trajectory. The g-forces continued to increase.

In Houston, everyone from Kraft and Charlesworth to the men in the farthest back rooms hardly breathed. With forty seconds to cutoff, Apollo 8 had reached about 98 percent of its target speed.

In Maui, spectators at an observatory watched the exhaust plume billow from the base of the Apollo 8 rocket as the engine burned through its last seconds of life. Inside the spacecraft, the crew heard almost nothing and felt little more than an ever-increasing *push-push-push* as the craft grew lighter with the burning of propellant.

"All right, fifteen coming up here," Borman called to his crewmates.

"Real fine. Ten seconds," Lovell affirmed.

"How's your inertial velocity?" said Anders.

"Velocity's looking fine," Lovell replied.

The men were now six seconds from leaving Earth.

"Five," Lovell called. "Four . . ."

The ship was traveling at more than 24,000 miles per hour. If the engine had one last push, it needed to push now.

No one moved at Mission Control.

The third-stage engine cut off. The astronauts and Mission Control looked at their instruments. Apollo 8 was 215 miles above Earth and traveling at 24,208 miles per hour—ideal for translunar injection and a trajectory to the Moon.

Immediately, Earth answered, using its gravity to pull back on the ship and slow Apollo 8's speed, but not nearly enough to overcome its momentum as it traveled away from the planet.

In Mission Control, Gene Kranz, who'd been a flight director for Apollo 7, got up from his seat, left the room, and broke down in tears at the magnitude of the moment.

Standing in the back row in Mission Control, Chris Kraft watched as the green blip on the screen moved away from Earth. When he'd joined the National Advisory Committee for Aeronautics, the precursor agency to NASA, in 1944, rockets were still the stuff of comic book covers and science fiction. Today, only twenty-four years later, he was watching men venture beyond their world—not on graph paper with theoretical trajectories, but in real time, in the sky above him. He felt the urge to speak, to say something to the astronauts, to announce something profound, but he was an old pro and didn't dare step on the CapCom's role. Rules were essential at a time and a place like this. The CapCom was a fellow astronaut—it was his reassuring, familiar voice the crew should hear during flight. To jump in now would break protocol and might cause confusion, and for that reason alone, Kraft believed it unethical to speak at all.

He watched the green blip move for a few more seconds, moving farther and farther from Earth.

Still, he needed to speak to someone, so he spoke to the green blip, and the entire room heard him.

"You're on your way," Kraft said. "You're really on your way now."

Chapter Thirteen

●

A DEEPLY TROUBLED YEAR

STREAKING AWAY FROM EARTH, THE ASTRONAUTS LEFT BEHIND a deeply troubled planet at the end of a deeply troubled year.

Nineteen sixty-eight had begun on an optimistic note, with a medical miracle. At Stanford University in California, Dr. Norman Shumway performed the first successful heart transplant in the United States. A week after the operation, fifty-four-year-old steelworker Mike Kasperak appeared to be doing fine.

As America picked up the confetti from its New Year's celebrations, the country's presidential campaigns began in earnest. President Lyndon Johnson, who'd won in a landslide in 1964, looked to be the certain nominee for the Democrats. The early front-runner for the Republicans was former vice president Richard Nixon, who'd begun to tour the country and make his case. In Alabama, former governor George Wallace was preparing a third-party run based on a pro-segregation platform that was popular in the deep South.

Looming over the campaigns was the war in Vietnam. Half a million American troops were in country; since 1965, when official combat units arrived, nearly twenty thousand American lives had been lost in the fighting against North Vietnamese Communist and guerrilla forces. Still, President Johnson and William Westmoreland, the general in charge of U.S. forces in Vietnam, promised the public that the war was going well and that victory was on the horizon.

On January 21, fifteen days after the operation, the heart transplant recipient in California died despite having made it through "a fantastic galaxy of complications," according to his surgeon.

Just after midnight on the final day of January, tens of thousands of North Vietnamese troops and Vietcong guerrillas launched a coordinated attack on nearly every major city and town in South Vietnam. The action came as a surprise to American troops, who were honoring a two-day cease-fire with the enemy during Tet, the country's sacred holiday. By sunrise, over 120 population centers and military bases had been assailed by more than 80,000 North Vietnamese and Vietcong fighters, an attack now being called the Tet Offensive.

For the first time, Americans were able to watch news coverage of combat without government control of images or information, thanks to reporters and cameramen embedded with the troops. The United States was supposed to be on its way to victory—the president and his generals had sworn to it—and yet here was an enemy that had stormed the American embassy and damaged nearly every stronghold in the south.

Night after night, the evening news showed graphic footage from the battle; often, 90 percent of the telecast was devoted to the war. One image sank especially deeply into the American psyche. In a still photograph and on film, it showed a North Vietnamese prisoner, hands tied behind his back, being executed by a single pistol shot to the head, delivered from a distance of a few inches by a South Vietnamese national police chief. There had been no charges, no trial, no last words—just the raising of the gun and a single shot to the temple. The photo ran on the front page of nearly every newspaper in America on the first day of February; no one who saw it, or watched the film of the shooting on the evening news, knew that the prisoner himself had executed, in cold blood, an entire family. All that America knew was that this terrible war was more ugly and brutal than they'd imagined, and that the clean and quick ending they had been promised seemed very far away.

In Orangeburg, South Carolina, a bowling alley remained one of the few local businesses to refuse service to black patrons, despite civil rights laws prohibiting such discrimination. In early February, black students at South Carolina State University began to protest, first by sitting at the lunch counter at the bowling alley, then by gathering in larger numbers and demonstrating outside. On February 8, a melee broke out during a student rally on the SCSU campus; panicked police fired into the crowd. The gunfire lasted just ten seconds or so, but when it was over, at least thirty people had been shot, and three black teenagers died. One of them, Delano Middleton, was a high school student whose mother worked on campus. At the hospital, Delano told her, "You've been a good mama, but I'm going to leave you now."

The next day, Governor Robert McNair called the episode "one of the saddest days in the history of South Carolina," but he blamed the violence on "Black Power advocates." At a time when hundreds of Americans were dying every week in Vietnam, the Orangeburg Massacre, as it would be called, soon faded from the headlines. But the future it foretold for 1968 was only just starting to crystallize.

During a background briefing ten days into the Tet Offensive, U.S. Secretary of State Dean Rusk erupted at reporters who were pressing him with tough questions. "Whose side are you on?"

Rusk demanded. The press was offended that Rusk would challenge their loyalties, but the reality was that the country was deeply divided about the war. Much of the difference in opinion fell along generational lines; older people tended to trust the government, younger people tended to question everything. (In fact, by 1968, a common expression among the counterculture was "Never trust anyone over thirty." And it was around that age that political opinions seemed to divide.) Thousands of roadside billboards admonished BEAUTIFY AMERICA, GET A HAIRCUT.

By late February, the Tet Offensive had ended. By all accounts, it was a resounding American military victory. Yet that was not the message delivered by Walter Cronkite to the nation during his February 27 newscast. The CBS anchor had traveled to Vietnam during the Tet Offensive to see things for himself. Cronkite rarely offered his opinion. Now, he spoke candidly, and viewers hung on every word:

"To say that we are mired in stalemate seems the only realistic, yet unsatisfactory, conclusion . . . it is increasingly clear to this reporter that the only rational way out then will be to negotiate, not as victors, but as an honorable people who lived up to their pledge to defend democracy, and did the best they could.

"This is Walter Cronkite. Good night."

When President Johnson saw the broadcast, he is said to have told those around him, "If I've lost Cronkite, I've lost the country."

On the same day that Cronkite addressed the nation, twenty-five-year-old Frankie Lymon was found dead on the floor of his grandmother's New York City apartment. Lymon had been a teenage singing sensation, part of the doo-wop group the Teenagers, and had been the angelic lead voice on songs like "Why Do Fools Fall in Love?" which he'd helped write at age thirteen. As his boyish voice deepened, he faded from public favor. Depression led to a heroin addiction, but by 1968, he claimed he was clean and hoped America would give him another chance. Police who found his body discovered a needle nearby. When news broke of his passing, people around the country pulled out their old Teenagers records and listened to Frankie ask, *"Why do birds sing so gay? And lovers await the break of day?"* To so many of them, the nineteen-fifties sounded like a very long time ago.

In 1967, President Johnson had appointed a commission, chaired by Illinois governor Otto Kerner, to study the race riots that had erupted in several American cities since 1965. On the last day of February 1968, the Kerner Commission issued its report, along with this conclusion: "Our nation is moving toward two societies, one black, one white—separate and unequal."

By March 1, after a month of news about the Tet Offensive, President Johnson's approval ratings had dropped by double digits. Infused with new energy, supporters of peace candidate Eugene McCarthy, a quiet, intellectual senator from Minnesota, campaigned almost nonstop in advance of the New Hampshire primary. Most of them were young and fervently antiwar; many even cut their hair and put on smart clothes to be "Clean for Gene." McCarthy was a long-shot candidate, but some thought he might get enough votes to avoid embarrassment.

On March 12, the day of the New Hampshire primary, McCarthy did much better than that. When the results came in, he'd tallied just 410 votes fewer than the president, a stunning near-upset. Immediately, New York senator Robert Kennedy announced that he would enter the race. Overnight, it seemed that an indestructible American president had turned to clay.

Nineteen days after the New Hampshire primary, President Johnson delivered an address to the nation on live television. At the end, he announced: "I shall not seek, and I will not accept, the nomination of my party for another term as your president." Shock waves rippled through the country. In nine months, America would have a new president.

In April 1968, Martin Luther King, Jr., went to Memphis, Tennessee, to support striking garbage workers. He led a march with black laborers, who held signs that read I AM A MAN. King, as always, intended to demonstrate peacefully, but the march turned violent. A black teenager was killed, and sixty protesters were injured. King himself had to be whisked away to safety.

Days later, he told a crowd at the city's Mason Temple Church: "Something is happening in Memphis, something is happening in our world."

He spoke of justice and fair treatment for the sanitation employees, of nonviolence, of the power of collective action. And he offered these parting words:

"Like anybody, I would like to live a long life. Longevity has its place. But I'm not concerned about that now. I just want to do God's will. And He's allowed me to go up to the mountain. And I've looked over. And I've seen the Promised Land. I may not get there with you. But I want you to know tonight, that we, as a people, will get to the Promised Land."

The next evening, April 4, at the Lorraine Motel in Memphis, King prepared for another rally. As he stood on the balcony outside his second-floor room, he called down to the parking lot to Ben Branch, a musician, and asked him to sing a special song for him at the event that night, an old spiritual, "Take My Hand, Precious Lord," one of King's favorites. A moment later, a bullet tore through King's jaw, severed his necktie, and ripped open his neck.

Ralph Abernathy, King's close friend and fellow civil rights leader, rushed to King and cradled him in his arms.

"Martin, Martin, it's going to be all right," Abernathy said.

An hour later, King was pronounced dead at St. Joseph's Hospital in Memphis.

That night, America started to burn.

Riots and violence broke out in 130 cities across the country. Over the next several days, 65,000 Army and National Guard troops were dispatched to try to keep the peace. In Chicago,

Mayor Richard Daley ordered his police superintendent to shoot to kill arsonists and shoot to maim or cripple looters. In Washington, D.C., soldiers stood guard on the White House lawn as fires raged just blocks away.

On the night King was shot, Robert Kennedy was scheduled to give a speech in Indianapolis. When news of the assassination reached city officials, they warned Kennedy that his safety could not be guaranteed. He went anyway. When he arrived at the corner of Seventeenth and Broadway, he climbed onto the back of a flatbed truck and stood out in the open. Wearing his late brother John's dark overcoat, he spoke to the mostly black crowd without looking at notes.

"For those of you who are black and are tempted to fill with—be filled with hatred and mistrust of the injustice of such an act, against all white people, I would only say that I can also feel in my own heart the same kind of feeling. I had a member of my family killed, but he was killed by a white man."

Never before had Kennedy spoken publicly about his brother's assassination. He called for compassion and asked for a prayer for the King family.

By the end of a week of rioting throughout the country, thirty-nine people had died, thousands had been injured and arrested, and millions of dollars of property damage had been done. Indianapolis, however, had stayed calm.

In mid-April, John Lennon and George Harrison became the last two Beatles to leave India after traveling there in February to study meditation with Maharishi Mahesh Yogi. They'd grown disenchanted with the spiritual leader after hearing he'd made a pass at women who'd joined their pilgrimage. When asked by the Maharishi why he was departing, Lennon said, "Well, if you're so cosmic, you'll know why."

Later that month, student activists occupied five buildings at Columbia University, took a dean hostage, and issued a series of demands. Among other things, the students insisted that the university end its association with a military think tank and halt its plan to build a gymnasium in Harlem on the site of a park used by lower-income residents.

Columbia officials resisted, only growing more entrenched as the students smashed furniture and shattered windows, destroyed academic research, and hung posters of Vladimir Lenin, Che Guevara, and Malcolm X on the walls. For a week, the university administration tried to wait out the protesters. Finally, they asked police to remove them.

At 2:20 A.M. on April 30, a thousand officers, many carrying flashlights and billy clubs, stormed the occupied buildings. Some students resisted passively, others by punching, biting, or throwing bottles and batteries. Many police officers used force, some of it brutal, to pull out the protesters and gain control. To some who watched, a class distinction could be seen in the collision of weathered boots with fresh faces, a working-class force smashing into private-school privilege.

The confrontation lasted past dawn. When it was over, more than seven hundred people had been arrested and nearly one hundred fifty injured, including twelve police. Shocked parents and other citizens looked at the photos of the aftermath and wondered what had become of their country.

Down Broadway from Columbia, a new musical was opening. *Hair* told a story of hippies, the antiwar movement, the counterculture, and the sexual revolution in 1960s America, and it featured drug references and group nudity. The sixty-seven-year-old reviewer John Chapman of the *Daily News* in New York called the show "vulgar, perverted, tasteless, cheap, cynical, offensive, and generally lousy" and recommended that "everybody connected with it should be washed in strong soap and hung up to dry in the sun." But even octogenarians who saw the musical had a hard time not singing along to the hit songs *Hair* produced, including "Aquarius" and "Let the Sunshine In."

In May, CBS television aired a special in prime time, *Hunger in America,* which told of the growing problem of malnutrition in the world's richest country. According to the documentary, there were ten million hungry people in the United States, a figure that stunned viewers. The filmmakers even showed footage of a dying, malnourished newborn. But perhaps the most memorable moments came in an exchange with a fourteen-year-old black student named Charles from Hale County, Alabama, who told a doctor he went hungry during the school day because he didn't have twenty-five cents to pay for lunch:

Dr. Wheeler: Well, what do you do while the other children are eating?

Charles: Just sit there.

Dr. Wheeler: How do you feel toward the other children who are eating when you don't have anything?

Charles: Be ashamed.

Dr. Wheeler: Are you ashamed?

Charles: Yes, they haunt you.

Dr. Wheeler: Why are you ashamed?

Charles: Because I don't have the money.

To win the Democratic nomination for president, Robert Kennedy had to win the California primary. A week earlier, he'd lost Oregon to McCarthy, and was trailing new entrant Vice President Hubert Humphrey in delegates. For RFK, the Golden State was the crossroads. If he lost there, he'd likely drop out.

As the California returns rolled in, it was clear Kennedy would win. Just before midnight, the candidate went to the sweltering ballroom at the Ambassador Hotel in Los Angeles and addressed a packed house of supporters. Looking more boyish than his forty-two years, Kennedy spoke of his belief that America could be healed and come together. In closing, he made a V with his raised fingers—which in 1968 stood for both peace and victory.

Followed by his entourage and a string of reporters, Kennedy made his way to the hotel's pantry, where he reached out to shake hands with Juan Romero, a seventeen-year-old busboy who'd delivered food to his room earlier that week. As the two moved close, a man with a pistol lunged forward, pointed the gun just inches from Kennedy, and began firing, hitting the senator once in the head and twice in the right armpit. As Kennedy collapsed, Romero cradled his head to protect it from the cold concrete and tried to comfort the senator, who had been kind to him a few days earlier and had treated him as an equal.

Photographers snapped photos of Romero holding Kennedy. The images would become among the most memorable of the twentieth century.

Pandemonium erupted throughout the hotel; supporters held their heads, sobbed, and screamed "No! No!" and "Not again!" Police seized the shooter, a twenty-four-year-old Jordanian American named Sirhan Bishara Sirhan. In his pocket they found a newspaper story noting Kennedy's support for Israel.

Kennedy was rushed to the hospital, where he clung to life. In England, the British Broadcasting Corporation told its audience, "We pray for the American people that they may come to their senses."

Early the next morning, on June 6, Kennedy died of his wounds. Across the country, people walked around dazed. In New York City, WPIX-TV broadcast the image of a single word—SHAME—and let it run for two and a half hours. People of all colors and classes and ages gathered spontaneously at railroad tracks to glimpse the train that carried Kennedy's body from New York City to Washington. When it passed, mothers holding babies waved, children saluted, the elderly tried to stand. Black and white Americans chased the train, running on the tracks together until the last car disappeared.

Richard Nixon became the Republican nominee for president on the first ballot at the party's national convention in Miami in early August. His running mate, Maryland governor Spiro Agnew, had backed the Civil Rights Movement but now scolded black people, and some of their leaders, for not disavowing so-called black racists. He and Nixon would run on a campaign of law and order, one aimed at a "silent majority" and voters shaken by the events of the year.

The Democratic National Convention was held in Chicago later in August. As protests erupted around the city, McCarthy seemed unwilling to seize the moment and lead the passionate supporters who'd backed him for so long. In the end, he received just 23 percent of the votes at the convention, and Johnson's vice president, Hubert Humphrey, who hadn't entered a single primary, became the Democratic nominee, with Senator Edmund Muskie of Maine his running mate. It was a crushing blow to believers in the antiwar movement.

In London, the Beatles were set to release a new single, "Hey Jude," written by Paul McCartney to reassure John Lennon's son, Julian, during his parents' divorce. It would include as its B side the John Lennon–penned song "Revolution," which questioned the tactics of the year's aggressive political protesters: "But when you talk about destruction / Don't you know that you can count me out." It had been just four years since the Beatles had dressed in matching suits and run from screaming fans in the opening scene of the film *A Hard Day's Night*. In 1968, they had long hair and flowing beards and were singing about the pain in the world. Even in music, little seemed the same anymore.

At the end of the summer, Atlantic City hosted the 1968 Miss

America pageant. Outside the venue, at least a hundred women protested the event, which they deemed exploitative. Officials had feared that the demonstrators would start fires, as happened so often during protests in 1968, but the women had promised they wouldn't do anything dangerous—"just a symbolic bra-burning." When the event began, they tossed false eyelashes and girdles into a garbage can and crowned a sheep Miss America. But from that day forward, the concept of a "bra-burning women's libber" gained currency in the United States, despite the fact that nothing was ever set ablaze.

In October, the Summer Olympic Games were held in Mexico City. Two American sprinters, Tommie Smith and John Carlos, won the gold and bronze medals, respectively, in the 200-meter race. Standing on the podium to receive their medals, each wearing a black glove on one hand, the two Americans bowed their heads and raised a fist in a Black Power salute during the playing of "The Star-Spangled Banner," a protest of the inequality of treatment and opportunity for black people in their home country. Immediately, the athletes were suspended from the team and sent home from Mexico City. The silent statement by the two sprinters had a polarizing and powerful effect in the United States.

On October 20, almost five years after her husband's death, Jacqueline Kennedy became Jackie O. Everyone seemed to have an opinion—mostly negative—of the surprise wedding between the former First Lady and the Greek shipping magnate Aristotle Onassis, a man twenty-three years her senior. It was a match far from Camelot. To many, it seemed that Kennedy had traded her quiet dignity and near-saintly virtue for a life with a crude, short, cigarette-smoking man who appeared to offer little more than money. And it seemed a break with a more innocent time, one when fairy-tale stories still happened in America.

As October bled into November, and with America just days away from electing a new leader, Jimi Hendrix's new cover of Bob Dylan's "All Along the Watchtower" rang out from cars and college campuses and protests, his guitar making a sound no guitar had ever quite made before, a black man singing a white man's song with the opening lyric, "There must be some kind of way out of here," and a reminder, in the midst of one of America's most volatile years, that "the hour is getting late."

Polling at 15 percent, independent presidential candidate George Wallace controlled millions of votes, most in the deep South. At a time when many in the country were offended by what they perceived to be a disregard for decorum and civility, Wallace made no secret of his contempt for the unkempt. "You got some folks out here who know a lot of four-letter words," he said when interrupted by hecklers. "But there are two four-letter words they don't know: W-O-R-K and S-O-A-P."

When the election results were tallied, Nixon received 43.4 percent of the vote, Humphrey 42.7 percent, and Wallace 13.5 percent. More than 73 million votes had been cast; Nixon's total exceeded Humphrey's by just 499,704—about the size of the population of Atlanta. For the first time in more than a century, a new president would not have his party control either the Senate or the House of Representatives. As with most everything in 1968, America seemed split in two.

In late November, the Beatles released their first double album. Officially titled *The Beatles,* it quickly came to be known as the White Album, for its stark white cover. It was worlds apart from their 1967 album, *Sgt. Pepper's Lonely Hearts Club Band.* Neither the songs nor the sides seemed connected, or to flow from one another; some of the lyrics were overtly political; and each member seemed, more than ever, to be writing solo material rather than as part of a whole.

Two weeks after the White Album's debut, the Rolling Stones released their own classic, *Beggars Banquet.* The LP featured the track "Sympathy for the Devil," sung by Mick Jagger from the perspective of Lucifer, asking "Who killed the Kennedys?" and answering "After all, it was you and me." In "Street Fighting Man," Jagger, who admired the spirit of revolution during 1968 and had even joined big protests, seemed to lament that the best help he could give was by singing.

As America entered the final two weeks of the year, a grim statistic emerged from Vietnam. More than sixteen thousand Americans had died in the war in 1968, by far the bloodiest year to date by a factor of almost 50 percent.

At their offices in New York City, the editors of *Time* magazine decided on its Man of the Year. Their criteria did not include virtue—only that the selected person be the one who most affected the news and embodied what was important about the year. Previously, the magazine had named luminaries such as John F. Kennedy, Winston Churchill, and Mahatma Gandhi.

For 1968, they chose THE DISSENTER.

Chapter Fourteen

●

A CRITICAL TEST

JUST EIGHT MINUTES AFTER THE THIRD-STAGE ENGINE CUTOFF, Apollo 8 burst through the altitude record of 853 miles set by Gemini 11 in 1966. But there was little time to celebrate, or even notice, the achievement. The spacecraft needed to separate from the spent third-stage booster, the S-IVB.

To make it happen, Borman turned a T-shaped handle that triggered a set of explosives, cutting loose the expired third stage in a spectacle of pyrotechnics. For five seconds after the separation, Borman slowly moved his ship away from the expired third stage.

Then he did something strange.

Using the thrusters, Borman pivoted the spacecraft 180 degrees and began to move back, nose first, toward the cast-off third stage. This maneuver wouldn't benefit Apollo 8, but it would be critical to future missions designed to land on the Moon. On those flights, the lunar module—the landing craft that would deliver astronauts to the Moon's surface—would be stowed on top of the third stage. (Apollo 8 carried a large cylindrical water tank to simulate the mass.) To retrieve the lunar module, the crew would need to return to it, pull it free from the third stage, and carry it to the Moon. Borman needed to prove that the maneuver would work. He couldn't dock, of course, but just needed to get close enough to confirm the procedure.

Now pointed back toward the third stage, the astronauts were awed by the view of Earth. Out the spacecraft's windows, the planet was a round swirl of vibrant blues and whites, with much of its curvature visible. Lovell grabbed a 70 mm Hasselblad camera mated to an 80 mm lens and began firing away, flying through a magazine of color film, capturing Cuba and Jamaica at the planet's bottom, never worrying about framing or filters lest he lose perspective.

Moments later, Lovell turned his camera to the still-glowing third stage. Hundreds of shimmering stars appeared to fill the sky, but the crew knew these to be fuel particles coming from the tanks or flakes of ice from the tank walls. In Hawaii, observers could see a speckled white fog in the sky as liquid hydrogen vented into the predawn dark.

At the speed Apollo 8 was traveling, more than 20,000 miles per hour, Earth appeared to shrink before Lovell's eyes, growing smaller with each passing second in the way a tunnel entrance appears to shrink to a passenger looking out a car's rear window.

"We see the Earth now, almost as a disk," Borman radioed to Houston.

"We have a beautiful view of Florida now," Lovell added. "We can see the Cape, just the point . . . and at the same time, we can see Africa. West Africa is beautiful. I can also see Gibraltar at the same time I'm looking at Florida."

Even Borman, who'd warned Anders not to spend time sightseeing, couldn't avert his gaze.

This must be what God sees, he thought.

Now it was time to move away from the third stage, which

had the same 20,000-mile-per-hour velocity as Apollo 8 and was following the ship into space—a wayward cylinder as tall as a six-story building, and twenty-two feet longer than the remaining spacecraft the astronauts were riding. Unlike the Saturn V's first two stages, which had fallen back to Earth by force of the planet's gravity, the third stage, like the spacecraft, had too much momentum for such a fate and continued to move along with Apollo 8 toward the Moon. If all went according to plan, Apollo 8 would pass just ahead of the Moon, while the third stage would pass its trailing hemisphere, then slingshot into orbit around the Sun.

But not all was going according to plan. The crew of Apollo 8 had lost sight of the third stage. Yet Borman knew it to be just one or two hundred yards away, a mere whisker in infinite space. And that meant trouble.

If the crew couldn't see the third stage, they couldn't be certain they wouldn't collide with it. For all the engineering miracles of the Apollo spacecraft, it hadn't been designed to absorb a ramming by an entire stage of a Saturn V rocket. The loss of visual contact came as a surprise to Borman; the crew hadn't trained for that. Anders could see that his commander believed an emergency to be unfolding.

Borman fired his thrusters and Apollo 8 gained one and a half feet per second in velocity—about one mile per hour. He could only hope the direction of the spacecraft's movement was away from the third stage, but he couldn't be sure because he still couldn't see it. For several moments, the crew could only sit there and wait for a possible impact.

Then, as if by magic, the third stage drifted back into view, just a little farther away than before.

Already, Apollo 8's evasive maneuver had altered its trajectory to the Moon. That could be corrected along the way, but the third stage still stayed with Apollo 8 like an old dog following its master. Borman estimated it to be just five hundred or a thousand feet away—much too close for comfort.

Borman could have fired his thrusters again, but in space, that had to be done in the correct direction, a difficult task to perform manually and by sight. If Borman added velocity in the wrong part of the spectrum of angles, he might only increase the spacecraft's vertical separation from the third stage rather than get ahead of it. Or he might even move Apollo 8 back into the third stage's path. That was one of the challenges of flying in formation in space; you had to know not just your own trajectory but that of the object you were trying to move toward or away from. And there was little about orbital mechanics that one could eyeball against a black background that stretched forever.

The spent stage moved even closer to the spacecraft.

From Houston, Collins recommended a separation maneuver.

"I don't want to do that," Borman answered. "I'll lose sight of the S-IVB."

"Frank, if you use zero, then make the [separation] if possible in the plus-X thrusters. That's the direction of the burn we'd like," Collins said.

"Well, can't do that. I'll thrust right square into that S-IVB," Borman argued.

For the next several minutes, Collins and controllers in Houston wrestled to determine a separation maneuver that would move Apollo 8 safely away from the third stage without causing Borman to lose sight of the threat. Making matters more difficult, the third stage continued to vent propellant, creating a virtual snowstorm in the jet-black sky. Earth was nowhere to be seen through the spacecraft's windows. Apollo 8 needed a landmark, some reference point against which to maneuver, but there was none in any direction.

Finally Houston devised a burn maneuver it believed could move Apollo 8 to safety. Borman didn't seem entirely comfortable with it, but he performed it using the spacecraft's thrusters, then waited to see where it took his ship.

"How is that booster looking now?" Collins called. "Is it drifting away rapidly, or how does it look?"

"We're well clear of the S-IVB now, Houston," Borman said.

While the astronauts had been focused on navigating away from the discarded third stage, Apollo 8 had passed through the Van Allen belts, two massive, doughnut-shaped bands of intense radiation that encircle Earth. Named for James Van Allen, the American space scientist who discovered them in the late 1950s, the belts had long been thought to pose a danger—even a deadly one—to space travelers. For years, scientists and government agencies had tried to figure out a safe way through the belts; Van Allen himself suggested that detonating a nuclear bomb might clear the ionizing particles enough to allow spaceship passage. In the end, NASA determined that the Apollo spacecraft would be traveling so fast, and the astronauts would be so well shielded by the command module, that the risk of harmful radiation exposure would be minimal.

But no one knew for sure.

To test it, NASA had fitted each member of the Apollo 8 crew with a Personal Radiation Dosimeter, an ivory-colored device about the size and shape of a bar of soap that Anders had helped to design. It would measure the levels of radiation to which the astronauts had been exposed. (The device had a tiny five-digit analog meter to provide readings.) Now that Apollo 8 had passed through both belts (the first of which extended from about 600 to 3,500 miles above Earth, the second from about 9,300 to 14,000 miles above Earth), Collins wanted a radiation reading from the crew. Borman was eager to learn the results; he'd heard the dire warnings some scientists had made about passage through the belts. Anders, a nuclear engineer who had developed expertise in shielding against charged space particles, felt certain no damage had been done. It was he who radioed back to Mission Control with the results.

"Houston. Apollo 8 with a PRD reading."

"Go ahead," Collins responded.

Anders gave the verdict. After passing through both belts, none of the astronauts had received more than about one-tenth the radiation of an average chest X ray. The command module was even better than a lead bib in protecting human beings from high-energy particles.

The spacecraft was now about 22,000 miles from home. Out the window, Earth had grown even smaller. The entire planet now fit in Lovell's center window.

"Good grief," Collins radioed, "that must be quite a view."

"Yes," Anders said. "Tell the people in Tierra del Fuego to put on their raincoats, looks like a storm is out there."

Apollo 8 was now five hours into its journey. In the two hours since it had left its parking orbit around Earth, its speed had decreased from more than 24,000 miles per hour to just 9,450 miles per hour as Earth's gravity continued to act on the unpropelled spacecraft. That decrease in speed would continue until Apollo 8 was about five-sixths of the way to the Moon, when lunar gravity would dominate and begin pulling the spacecraft toward its surface, causing the speed to rise again as the astronauts fell toward their target.

But even at these decreasing speeds, Earth continued to appear smaller every time the crew looked back. To Anders, it felt like watching the clock in fifth grade: If you stared, it didn't seem to move, but if you looked away and then looked back a short time later, it had changed.

If Apollo 8 were allowed to fly freely now, without any midcourse corrections to its trajectory, it would coast for about three days, then smash into the Moon. Midcourse adjustments would be necessary, as many as four, if needed. But those would come later.

Around six and a half hours into the flight, the first shift change occurred at Mission Control, when the Maroon Team, led by Flight Director Milt Windler and CapCom Ken Mattingly, took over from the Green Team. They would run the flight for the next eight hours until their replacements, the Black Team, took over. After that, it would be back to the Green Team, and so on.

Thirty minutes later, Apollo 8 prepared for another of its critical maneuvers. Until now, the spacecraft had flown with one of its sides exposed to the Sun and the other side facing away. But that arrangement couldn't last much longer without damaging the ship by broiling one side and freezing the other. To solve the problem, NASA had developed a procedure called passive thermal control, in which the commander slowly rotated the ship on its long axis, making one full revolution every hour as the craft journeyed through space. In that way, temperatures would become evenly distributed as the spacecraft turned on its invisible rotisserie spit. The maneuver had earned the nickname "barbecue mode" at NASA, and as Borman tapped a thruster, Apollo 8 became the first to use it in space. So slow was the roll that the crew hardly sensed it.

Now, as they continued to streak toward the Moon, the astronauts prepared for emancipation. More than eight hours and 45,000 miles into their mission, the time had come for the men to slip out of their bulky space suits. They had been wearing them since long before launch in order to breathe pure oxygen. Doing so helped to purge nitrogen from their bodies, and that had been critical during launch. As expected, the cabin's internal air pressure dropped rapidly as the spacecraft ascended. If the crew had not purged the nitrogen from their systems, the sudden drop in pressure could have caused the nitrogen to form bubbles in their tissues that could press on nerves, lungs, spine, even the brain—a painful and potentially deadly condition common to deep-water scuba divers who surface too quickly, and known as the bends. Once in space, the astronauts had been so busy with equipment checks and the translunar injection burn that they hadn't had a chance to remove the suits. Until now.

The crew doffed their suits and stowed them in bags under their seats, where they would remain for the duration of the flight. Suddenly unbound, Anders was free to test his new super-power: weightlessness.

A breath of a touch, just enough to light an elevator button, propelled him to any destination inside the cabin. If there were twists and turns in his way, he simply bent or hunched or cork-screwed to conform to the openings, flowing through them like water. Using his fingertips for thrusters, Anders visited naviga-tion instruments, storage areas, labyrinths of valves. His light-weight coveralls, made of fire-resistant Teflon beta cloth, were perfectly white except for a large American flag patch sewn onto the right shoulder, the Lovell-designed figure eight mission patch on his left shoulder, and a red and blue NASA patch on his right breast. They made for a perfect flying skin, cinched at the waist and tight enough to keep his body slender.

Like electricity moving through a circuit board, Anders made turns and changed directions with precision. Sometimes he looked like a giant strand of spaghetti being slurped through the cabin, yet he managed never to bang into anything or catch any of his coveralls' pocket pouches on equipment. When he stopped moving, he put objects into twirling, spinning, tum-bling motion—a penlight, his camera, even the chewing gum in his mouth all came to life in a floating symphony that Anders conducted. Soon he got into the act himself, doing a somersault, then another. As he got ready for a third, he thought, *Whoa. What's that hairy feeling coming up in my throat?*

And he knew—if he flipped again, he would vomit.

Slowly he floated back to his seat, strapped himself in, and stared at the instrument panel, trying to orient himself in a world that had suddenly lost its up and down.

Although the ship still flew "backward," blunt end toward the Moon, the crew constantly had new views out their windows, courtesy of barbecue mode. By now, Earth looked to be the size of a softball. Its beauty made Anders forget his roiling stomach.

He could pick out the distinctive shape and deep blue waters of the Tongue of the Ocean in the Bahamas, and the iridescent turquoise of the shallow waters that framed it. He could see the horn of Africa in its sunbaked salmons and browns. But sometimes there were so many clouds it was hard to distinguish what was what. The nun who'd taught Anders's first grade class had kept a globe on her desk, and it had the North Pole at the top. That's how every globe Anders ever saw had looked, with the world positioned logically, north on top. And that's how he sought to make order of things out the spacecraft window. Yet sometimes the views made no sense—things weren't where they were supposed to be, shapes didn't belong. And what was that giant white spot on the top of the planet? For several moments, Anders couldn't place it. Then it hit him: Antarctica—the continent that was supposed to be at the bottom of the globe. Even now, he had to remind himself that there was no up or down in space, or even in the universe. It had always been a matter of perspective, and now he had to change his. Turning himself upside down, he now understood Antarctica, the hemispheres, and the shape of a world he'd seen only one way until now.

While the astronauts continued to move through their checklists and monitor the spacecraft and its systems, they also had to tend to personal systems. Going to the bathroom was a challenge, largely because the command module had no dedicated facilities.

Urinating was a straightforward, if inelegant, process. It began with a kind of open-ended condom for which the astronauts had been fitted during training. (The devices came in small, medium, and large sizes, but astronauts assigned a more scientific nomenclature to the fittings: "extra-large," "immense," and "unbelievable.") Once out of his suit, the crewman would slip on the condom, then belly up to a valve and attach the other end to a bypass valve that, once it was opened, vented out the side of the spacecraft. If the procedure was timed properly, the astronaut could open the valve while urinating and expel the waste into space. If it was not timed properly, he risked exposing his tender parts to vacuum forces. To prevent that, Anders opened the valve too slowly on his first attempt, blowing off the personal end of his condom and sending twinkling golden droplets dancing weightless through the cabin. His timing improved after that, he made sure of it.

Even when urine was expelled properly from the spacecraft, the crew couldn't quite be done worrying about it. Just the tiny force necessary to vent the liquid—which turned to gleaming ice crystals in the sunlit cold of space—could have a profound effect on the spacecraft's trajectory and would have to be accounted for as the ship continued its journey.

Defecation was even less glamorous. The astronaut started with a collection bag fitted with an adhesive collar. After stripping naked (usually in private at the other end of the command module), he pressed the collar around his hind end until it stuck, then expelled to the best of his ability. In space, clumps didn't drop from the body. To help that along, NASA had built a narrow pouch into the bag for the astronaut's finger, which he could use as a scooper to pull things free. Finally, a packet of blue germicide was deposited in the bag, then ruptured and kneaded together with the waste in order to neutralize odors and to kill bacteria that could, over time, generate gases that could cause the package to explode. This bathroom breaks could take as long as an hour. Cleaning was done with a small moist towelette like those handed out at barbecue restaurants.

Much as the crew might have liked to fire the sealed bag into space, they could not. Ejecting such a bulky item would require the cabin to be depressurized, possible but risky to the men and the flight. Also, NASA planned to examine the feces (as well as blood and urine) on the crew's return to Earth, eager to study the effects of deep space flight on the human body.

Even as an engineer, Anders knew this fecal collection system would be difficult. Months before Apollo 8, he took home a kit to practice (one didn't experiment on such a device in the simulators at work). He explained to Valerie that it had to be tested, at least on Earth, while lying down. To that end, he intended to try it in bed.

"Not in our bed!" Valerie said.

So Anders lay on the carpet and gave it his best.

The device did not work well for him.

A few days later, he asked the flight surgeon to recommend a low-residue diet he could eat in the days leading up to and during the flight. The less often he had to use the device on the mission, he figured, the better. So far, his plan was working. While Borman and Lovell struggled with the contraption, Anders sat in his seat, doing his work and looking out his window, uncalled by that part of nature, watching the universe go by.

More than nine hours had elapsed in the flight before the astronauts got their first glimpse of the Moon. It happened during one of Lovell's looks through the spacecraft's telescope and sextant, when he spotted a barely visible crescent surrounded by a light blue haze, "just about as light blue as we have it back on Earth," Lovell radioed to Houston. Lovell knew that the appearance of color came from the way the Sun scattered light through his navigation instruments. Tiny as the Moon appeared through his telescope, it was more than his colleagues were getting, or were likely to get, given how the spacecraft (which had just five small windows) needed to be oriented during flight. The Moon was still sixty hours away, and any good views available to the crew might have to wait until they arrived.

Around nine and a half hours into the flight, Apollo 8 was more than 52,000 miles from Earth and weighed just 63,295 pounds—less than 1 percent of its launch weight, thanks to all of the spent propellant and discarded stages. Soon it would weigh a bit less than that, as the crew got ready to do something many at NASA thought it should not do.

To enter and exit lunar orbit, Apollo 8 would rely on its Service Propulsion System, a single engine designed to slow down or speed up the spacecraft as needed. It was one of the most important pieces of equipment on the vehicle, and the only major source of propulsion remaining on Apollo 8. If it did not function properly, the results could be catastrophic for the crew.

Engineers and controllers had confidence in the SPS engine, because it had been tested repeatedly on the ground and had worked well aboard Apollo 7 just two months earlier. But to Chris Kraft, that wasn't enough. Despite its track record, the engine had never been tested in deep space. And it was, after all, a rocket, and rocket engines were complicated, and never one hundred percent propositions.

So when the flight plan was devised for Apollo 8, Kraft wanted the engine tested in flight, a confidence burn for just a few seconds, when the spacecraft was about 60,000 miles from Earth. That way, if anything malfunctioned, either with the engine or with the computers that ran it, the problems could be fixed or the flight aborted.

Some controllers protested Kraft's proposal. Test-firing the SPS, they argued, could screw up Apollo 8's trajectory, fouling its path to the Moon. But Kraft wouldn't back down.

"Fire that thing and I'll get it back on trajectory for you," he assured them. "But I want that engine to run before we get to the Moon."

And that was how the flight plan stood until about an hour before the test was to run.

Now controllers doubled down on their concern that the spacecraft would be thrown dangerously off course by the two-and-a-half-second burn. They urged Kraft to abandon the plan. Doing so would have been highly unorthodox—NASA's practice was to preplan flights down to the minutest detail in order to avoid surprises and unknowns. Kraft held firm, and he was the boss.

"We're going to do it, so let's do it," he said.

The astronauts spent the next hour preparing for the burn.

Inside the spacecraft, Lovell was singing random songs to himself. (Borman was accustomed to Lovell's singing from their time together aboard Gemini 7; for his part, Anders seemed too busy to notice.) When the moment came, Borman began his countdown. At five seconds, the computer's display flashed 99—a request to go ahead with the test. Lovell reached forward and pressed the Proceed key.

The SPS engine lit, and the crew felt a gentle push forward as the spacecraft gained speed. After 2.4 seconds, the engine cut off, just as programmed.

A few minutes later, the Public Affairs Officer made a happy announcement.

"The burn was completely nominal in all respects."

And for a minute or two, even Kraft believed it.

Chapter Fifteen

●

AN ASTRONAUT IN TROUBLE

IT WAS SATURDAY EVENING IN HOUSTON, ELEVEN HOURS AFTER launch, and Valerie Anders needed to get out of her house. With squawk boxes still chirping, she asked the family's au pair to watch her five children, then slid out the back door, careful not to make a sound lest the swarm of media camped out on her front lawn discover the subterfuge and follow her on her mission.

She arrived at her destination about ten seconds later—the neighboring home of astronaut Charlie Duke, whose family was hosting a Christmas party that night. The eggnog was flowing, and best of all, CapComs Mike Collins and Jerry Carr were in attendance, and they gave Valerie the skinny on the flight. Everything, they assured her, looked smooth so far.

Five miles away, at Mission Control, Chris Kraft was pacing.

The SPS engine had not worked properly.

Thrust buildup had been slow, and overall thrust had been

too low. These were results NASA had never seen in tests, and they presented big problems for Apollo 8. An explanation, and a fix, had to be found soon, or lunar orbit might be impossible. One thing Kraft did not intend to do was inform the crew, at least not yet. There was nothing the astronauts could do about it anyway.

Several minutes later, Borman turned in for the first sleep shift of the flight. He was scheduled for seven hours, after which Lovell and Anders would rest. In Houston, engineers were studying the poor results from the brief burn of the SPS engine. No one had a clue as to a cause or a fix. Despite its being a Saturday evening, and right before Christmas, experts flowed into Mission Control to analyze the problem as Apollo 8 continued with its lunar intentions.

At her home, Susan Borman kept entertaining the parade of guests who came to support her, always keeping one ear trained on the squawk box, nodding in conversation without actually hearing what people were saying. Her favorite dialogue between Apollo 8 and Mission Control came when Frank said things like "We noticed on our system test battery vent pressure that when we opened the battery vent valve, we get an immediate drop-off to pressure which nulls out at about two-tenths to three-tenths of a volt"—not because she understood the jargon, but because the sound of his voice proved Frank was still alive.

Down the road, Chris Kraft, Flight Director Glynn Lunney, and several mechanical minds were studying 2.4 seconds' worth of data, trying to explain the loss in pressure and thrust in the SPS engine. After nearly two hours of frenetic analysis, a contractor from North American Aviation, which built the spacecraft, had an epiphany: A bubble in a propellant line had fouled things up.

Helium, the man reasoned, must have become trapped during launch and remained in the oxidizer line. That's why the engine didn't achieve full thrust right away. One could hear the same thing when starting a lawn mower after a period of inactivity. If that was true, it was good news for NASA, because it meant the bubble likely had been purged and the flow of propellant purified. But no one could know for sure. It was now up to Lunney and Kraft to decide what to do with that theory.

As they mulled over how to proceed, Borman radioed Houston. He was supposed to be sleeping but, in the excitement of the flight, couldn't make it happen.

"We have one request. CDR would like to get clearance to take a Seconal."

Borman had asked whether he (CDR was shorthand for Commander) could take a sleeping pill. He detested the idea of relying on medication, but it was almost impossible to shut down one's brain in the middle of mankind's first trip to the Moon. Borman figured that a single Seconal, a barbiturate often prescribed for sleep, wouldn't be harmful under the circumstances.

CapCom Mattingly checked with NASA's doctors, who okayed it, and Borman made his way back down to the sleeping area in the navigation bay. His crewmates were working and talking above him, but it was the best refuge possible in a craft just thirteen feet by eleven feet and filled with equipment.

After much discussion, Lunney and Kraft came to a decision on the matter of the SPS engine. Each of them thought through the theory offered by the man from North American and ultimately judged it to be correct, and they were willing to bet the rest of the flight on it. After consulting with other controllers, who concurred, the men decided to continue with Apollo 8's flight plan just as it had been written. The next time the SPS engine fired would be when the spacecraft slipped behind the Moon. At that moment, the crew's lives would depend on its functioning properly.

As midnight approached in Houston, all three of the astronauts' wives, too, struggled to sleep. It had been a long and exhausting first day, but Susan and Valerie remained attached to their squawk boxes as they lay in bed, each trying to pick out a hint of how her husband was feeling by the tone of his voice.

Staying overnight in Florida, Marilyn Lovell had no squawk box; instead, she listened to the sound of the waves by her beachside cottage, wondering whether Jim could see that same stretch of ocean from space.

A few hours after test-firing the SPS engine, Houston got good news. The test burn hadn't fouled Apollo 8's trajectory, as some had feared. In fact, tracking analysis showed that if the spacecraft made no further changes and was simply allowed to coast, it would slingshot around the Moon at an altitude of just 80 miles above the lunar surface, then return to Earth, just as the trajectory specialists had designed.

For the first time in a long while, Mission Control grew quiet. It was past midnight and the spacecraft was coasting. And Borman was supposed to be sleeping.

Instead, he tossed and turned in his hammock. Borman had never been sick for a minute on the two-week flight of Gemini 7, or even on the "Vomit Comet," the zero gravity airplane used to acclimate astronauts to weightlessness. Even when flying in violent thunderstorms as an inexperienced fifteen-year-old student of Miss Bobbie Kroll, he'd not experienced so much as a stomachache.

Now he swallowed hard in his sleeping bag and tried to push away the nausea, but the waves were building and moving fast toward shore.

"I'm sorry, guys," he called to his crewmates above.

And then the vomit came.

Retching, Borman reached to capture the floating green globules, but there were too many of them, going in too many directions, to corral at once. Even when he caught them, they just split in two or four or eight and made their escape from his flailing hands.

A moment later, the odor of the vomit reached Borman's two crewmates. Overwhelmed, Anders reached for his gas mask.

"You're not supposed to use those!" Lovell said.

"To hell with that, I'm using it," Anders replied. He opened the oxygen supply to maximum, then turned his attention to Borman.

From below, he could see a greenish-brown blob, about the size of a golf ball, moving toward him. For a moment, the physicist in him took over, and Anders followed the object with wonder as it oscillated in three dimensions, a movement impossible on Earth, and quivered toward the ceiling. Anders's instinct was to find a camera and photograph the alien wonder, but he couldn't tear his eyes away as it rose higher and then, about eighteen inches from his chest, split like the atoms he'd seen in science films, one wobbling part headed this way, the other wobbling in the perfect opposite direction. Anders thought, *That's Isaac Newton. That's conservation of momentum.* Now one of the pieces was heading toward Lovell, who could do no more than watch it, eyes narrowing as it hit him in the chest and spread like an uncooked egg against the white cloth of his coveralls.

Lovell reached for a towelette and tried to wipe the mess away, but his and Anders's troubles were only starting. Floating toward them from below were spinning blobs of feces, each turning on its own axis. If they had been solid clumps, Lovell and Anders might have had a chance to dodge or capture them, but Borman had diarrhea.

Lovell and Anders grabbed as many wipes as they could find and began hunting down the fluttering pieces, netting them like butterflies. For several minutes, the three men worked to clean the cabin. After restoring some order, Lovell and Anders could see that Borman was very sick. The situation, Anders thought, needed to be reported to Houston right away.

"Absolutely not," Borman said.

Anders understood his reaction. Borman was a test pilot in his bones; no one with his instincts or credentials would want the world to know he'd become sick in space. And Anders didn't blame him—he would have felt the same way himself.

But it was more than that to Borman. He didn't trust NASA's

doctors, especially the agency's medical director, Charles Berry, whose judgment he questioned and who he believed to be ever itching to make himself part of the story. And it wasn't just Dr. Berry who worried Borman. Give any NASA doctor a chance to play the hero, he believed, and you were asking for trouble. Borman could imagine it happening now, some medical guy stepping in and canceling the mission "for the good of the crew." Borman would rather have died than foul up Apollo 8. News of his illness would remain between him and his crew.

Lovell agreed. He saw NASA's doctors in much the way Borman did—eager to become major cogs in the wheel of space exploration. He remembered how he'd been rejected on his first application to the astronaut corps on account of a slightly elevated level of bilirubin, a phony excuse if ever there was one. If Borman was too sick to continue, Lovell thought, he and Anders would feed their commander, watch him, take care of him, and finish the mission. What they couldn't afford now, as they drew closer to the Moon, was to be ordered by Houston to turn back.

Anders wasn't so sure. What if Borman didn't get better? What if he got so sick that he and Lovell had to focus entirely on taking care of him, and none of the crew could work? But he could see that it didn't matter. This wasn't a request from Borman; it was an order. Lovell and Anders were military men. They understood the chain of command. And Borman, sick and covered in unpleasantness as he was, remained the commander. So nobody said a word to Houston.

Still, Anders knew the flight was at risk. Whether NASA knew about Borman's illness or not, there was no way Apollo 8 could go into orbit around the Moon with a guy who was vomiting and had diarrhea—and who might be getting worse.

Foul-smelling miniglobules of vomit and feces speckled the cabin as Borman fought his sickness and took control of the spacecraft while Lovell and Anders tried to sleep. Despite Borman's illness, his crewmates had faith the old fighter pilot could handle the spacecraft. Out his window, Borman watched Earth, now a hundred thousand miles away, grow so small it fit behind his outstretched fist.

Anders climbed into his hammock, which he found too big for his small frame. Worse, he had nothing to cuddle up to—no extra pillow, no covers, no Valerie. Whenever he dropped off to sleep, he suddenly felt like he was falling, as people do in dreams, and it would jar him back to consciousness. He tied a knot in his sleeping bag—something to press against—and that helped prevent the feeling of falling, but when he drifted off to sleep, the primitive level of his brain shouted "What the hell is going on here? Where are you?" and he'd wake again.

Lovell couldn't sleep either. Every few minutes, he saw tiny novas of white light exploding in his field of vision, which was odd since his eyes were shut. Holding his hands over his face did nothing to stop the fireworks. Lovell didn't know it, but his optic nerve was being bombarded by (mostly harmless) cosmic rays. Shielding his face did nothing, as the rays just danced on through. And if either man was lucky enough to find a few minutes' sleep, intermittent radio noise would shake him from it.

Twenty-four hours into the flight, Apollo 8 crossed the halfway point on its journey to the Moon. Owing to its continued decrease in velocity since leaving parking orbit around Earth, the spacecraft was traveling more slowly now, and it was still forty-five hours from its destination.

Soon Lovell and Anders were awake and back in their seats. To both of them, Borman looked a bit healthier, a condition the commander confirmed.

"I think it was the sleeping pill," Borman said.

But Anders wasn't so sure. A Seconal might explain vomiting, but not diarrhea. He still thought Borman should report that he'd been ill. There was a chance the commander was still sick and might get worse. And there might be an easy antidote, if only they would give Mission Control the chance to suggest one.

By now, however, Anders realized that even if Borman had been willing to report his condition, the media would hear the broadcast (as they did the vast majority of them) and jump to its usual worst-case conclusions, and a public relations nightmare would ensue. That wouldn't be good for anyone—not Borman, not their wives, not NASA. So Anders pitched another idea.

Borman could make a tape recording that described his illness, which then could be sent to Houston via an auxiliary channel meant for television, data, and backup voice transmissions. Only a select handful at Mission Control would hear it.

Borman thought it over. He still believed the doctors would leap at the chance to insinuate themselves into the flight. But Anders was right, maybe there was a fix, or at least something to learn, if NASA was informed in a discreet fashion.

"Okay," Borman said. "Let's do it."

Anders engaged the spacecraft's voice recorder with its built-in voice track, and ensured that there were no downlink communications to Houston. Then Borman started talking, and he spared no details. A few minutes later, Anders pressed some buttons and shot the recording back to Earth.

The crew expected to hear back from Houston in a few minutes. Instead, two hours passed without mention of Borman's condition by Mission Control. Anders snapped a few photos of Earth, wondering what could be taking Houston so long. Finally, Collins radioed to the spacecraft and said that Houston had received the voice tape and would advise shortly.

When Mission Control listened to the tape and realized what Borman was describing, top management rushed together, including the flight directors, Chris Kraft, and Dr. Berry. Kraft was furious that Borman hadn't reported his illness right away. While he understood that astronauts didn't trust doctors or want them mucking around in their domain, he couldn't abide the test pilot ethos of silence; Mission Control was there to assist the crew. But if no one talked, it rendered Houston helpless.

But Kraft didn't have the luxury of frustration now. He and

the others had to figure out what had happened to Borman in order to determine what to do about it—and what to do with the flight.

Dr. Berry considered that Borman might be suffering from a virus, perhaps even the Hong Kong flu that had struck so many in recent weeks. That was the fate that had worried Valerie Anders when she saw guests coughing and sneezing at the White House during the astronauts' last-minute send-off. But the most ominous possibility was also the simplest: that there might be something NASA and doctors did not know or understand about humans going to the Moon. In that case, it would be hard even to guess at a remedy, if there was a remedy at all.

Despite their uncertainty, the managers and Dr. Berry had to make a decision. If Borman had been made sick by radiation, a virus, or some unknown cause related to lunar travel, it was likely his crewmates would become sick, too. It would be difficult enough to justify continuing the mission with one astronaut in trouble. To continue with two or three out of commission was unthinkable.

The decision makers began to discuss aborting the mission and returning the crew.

On board Apollo 8, Borman, Lovell, and Anders awaited the verdict.

An hour later, the call came in.

The bosses had made up their minds.

Chapter Sixteen

●

EQUIGRAVISPHERE

DURING SUNDAY MORNING SERVICES ACROSS AMERICA, CON-
gregations prayed for the astronauts. In Rome, Pope Paul VI did
the same: "We open the window and instinctively the eye, the
thought, the heart, go to the heavens. We pray to the Lord for
them, and for the world, which is dazed at the conquest of science and of human endeavor." Leaving St. Christopher's Episcopal Church in League City, Texas, on the arms of her two big,
rugged teenage sons, Susan Borman remained grateful for everyone's good wishes, even as she calculated that Frank had moved
another ten thousand miles away from her since services had
begun.

At home, Susan climbed out of the family's old F-150 pickup
truck. Reporters were waiting on her front lawn, and she smiled
and answered questions, then excused herself and went inside.
Ignoring all the food left by well-wishers, she made her way to

the bedroom, where she turned off the lights, lay on the bed, and listened on the squawk box for the voice of her husband.

A few miles away, Marilyn Lovell and her four children had returned to Houston from Florida. When she opened the door to her house, she was greeted by a small village of friends, baby-sitters, neighbors, and astronauts with their wives, all of whom had brought something to eat or to drink (including the customary deviled eggs and champagne). The first thing Marilyn did was go to each of the four squawk boxes set up in her home—in the study, master bedroom, family room, and living room. Only after she'd flipped each of them to ON did she circle back to join her company. (Both Marilyn and Susan were squawk box veterans, having listened in during their husbands' flight together on Gemini 7.) Neither Marilyn nor the other astronaut wives understood much of the technical jargon, but all of them found comfort in hearing their husband's voice and those of the men they knew in Mission Control.

Now, however, when Mission Control called the spacecraft with their verdict on Borman's illness, none of the wives could hear it, nor could the media.

"We are on a private loop now," Collins said to the crew, "and we would like to get some amplifying details on your medical problems. Could you go back to the beginning and give us a brief recap, please?"

"Mike, this is Frank. I'm feeling a lot better now," Borman responded. "I think I had a case of the twenty-four-hour flu."

Given that flu viruses could be contagious, that might not have been the best answer to provide.

Collins asked the commander to review the history of his illness—when it started, what symptoms he experienced, the works. Borman provided the details. Then Dr. Berry jumped on the line—the man Borman least wanted involved.

"Frank, this is Chuck. The story we got from the tape . . . went like this: At some ten to eleven hours ago, you had a loose

BM, you vomited twice, you have a headache, you've had some chills, and they thought you had a fever. Is that affirm?"

"Everything is true, but I don't have a fever now. I slept for a couple hours and the nausea is gone, and controlling the loose BM. I think everything's in good shape now."

"Did you have a sore throat?"

"The roof of my mouth was sore, roger."

"And as we understand it at the moment, Frank, neither Bill nor Jim have anything at the present time except some nausea. Is that right?"

"No, none of us are nauseated now. We're all fine now."

Dr. Berry told Borman to take a Lomotil tablet, an antidiarrhetic. If needed, Borman was also to take Marezine, a drug used to counter nausea, vomiting, and other symptoms associated with motion sickness.

But Borman and his crewmates were more interested in what Berry had not told them to do.

He had not told them to come home. At least not yet.

Dr. Berry looked at Kraft, the flight directors, and General Samuel Phillips, director of NASA's Apollo Manned Lunar Landing Program. Ultimately, the decision belonged to Flight Director Cliff Charlesworth, but everyone was involved in this determination. Based on what they'd all just heard, a decision had to be made now, on the spot, about aborting the mission.

The men spoke for a few moments, then motioned to Collins to radio back to the crew.

"Apollo 8, Houston," Collins called. "We are closing this circuit down and we will be up in our normal voice loop in about five minutes. And then we will get on with the water dump."

By which NASA meant, "Let's keep this thing going."

Dr. Berry and the others had determined that Borman's illness had passed, and that if it recurred, it could be treated. Not long after, the doctor explained his thinking in a press conference, telling reporters that "this may be the type of thing that we see with motion sickness, it is just going to take some more watching to see."

NASA's public affairs officer announced the same to America. Listening at home, Fred Borman could only smile. He knew his father. Even if he'd suffered a heart attack and was lying paralyzed in the spacecraft, he would have ordered Lovell and Anders to continue the mission. That was his dad.

Apollo 8 was now 140,000 miles from Earth and just 100,000 miles from the Moon. In about an hour, the crew would be making its first live television broadcast. It had been more than twelve hours since Borman had taken sick. Now he felt better.

As the telecast time drew closer, the spacecraft's high gain antenna was adjusted and communications checked. The antenna, comprising four 31-inch dishes, could swivel to point at Earth to send and receive tracking, voice, and television signals. When the astronauts of Apollo 7 had made their appearance in living rooms across America two months earlier, they had done so from an altitude of about 150 miles. When Apollo 8 would go live for the first time, it would do so from almost a thousand times that distance. No one knew if it would work.

Barbecue mode was halted so that the antenna could remain pointed at Earth. In the command module, the crew worked to set up a four-and-a-half-pound black-and-white RCA video camera fitted with one of two available lenses—one to show the inside of the cabin, the other to show the views out the window. If all went well, the broadcast would begin at about three in the afternoon Eastern Standard Time in the United States, when many families would be home watching Sunday's professional football games. Borman hadn't wanted to bring television cameras in the first place, and when the flight plan was being made, he had bristled at the idea of interrupting NFL playoff action, which he would now be watching himself if only the high gain antenna could pull in the signal from Earth.

Before the scheduled broadcast, Valerie had gathered her children in front of the family's television and flipped on the special programming, then gone out to answer a few questions from reporters. When she returned, the TV was tuned to cartoons, a situation she quickly remedied. The broadcast began a minute later.

"Are you receiving television now?" Borman asked Houston.

On millions of sets across America, a gray screen flickered and flashed. Suddenly there was Borman, slightly blurry, diagonal, and seated at the controls of Apollo 8, his right hand on a joystick-shaped thruster control, his left hand waving to the world.

"Okay," Borman said, moving the thruster, "we're rolling around to a good view of the Earth, and as soon as we get to the good view of the Earth we'll stop and let you look out the window at the scene that we see. Jim Lovell's down in the Lower Equipment Bay preparing lunch, and Bill is holding a camera here for us both."

Anders swung around for a view of Lovell, who was working upside down. A bag floated in the cabin nearby. Borman continued to swivel the spacecraft with the rotation thrusters.

"Okay, now we are coming up on the view we really want you to see, that's the view of the Earth, and if you'll break for just a minute, Bill's going to put on the large lens. So we'll be right back with you."

A few moments passed as Anders changed lenses and repositioned the camera. His job was made tougher by the fact that he had no monitor to show him what he was capturing—this was strictly a point-and-hope affair.

"Houston, we are now showing you a view of the Earth through the telephoto lens," Borman announced.

But viewers saw nothing but a test pattern of vertical gray bars. For nearly four minutes, Anders and Borman wrestled with lenses and settings. Finally, an image emerged of a bright round object out the window—Earth!—but to viewers it looked featureless and indistinct, more like the Sun. Something was preventing the telephoto lens from getting the shot.

Borman switched back to a shot of the cabin, where Lovell grabbed a bag of chocolate pudding that was floating by. Nearby, Anders made his toothbrush dance and tried catching it with his teeth. Borman, who'd argued against these broadcasts, now couldn't hide his disappointment at being unable to share his breathtaking view of the world with the world.

"I certainly wish we could show you the Earth," he said. "It is a beautiful, beautiful view, with predominantly blue background and just huge covers of white clouds . . ."

At their homes, the astronauts' wives wished they could see more of their husbands, or even just a little color in their faces. Ten-year-old Glen Anders thought his floating father looked weird.

Anders moved in for a close-up shot of Lovell.

"Bill, you can let everyone see he has already outdistanced us in the beard race," Borman said. "Jim has got quite a beard going already."

Lovell turned to the camera with a big smile.

"Happy birthday, Mother!" he said.

Blanch Lovell had turned seventy-three that day, and Lovell hadn't forgotten. Watching on television at home in Edgewater, Florida, Blanch was stunned that Jim would remember her birthday at a time like this.

A few seconds later, Borman told his audience he needed to put his ship back into barbecue mode in order to prevent overheating. There was nothing he could do to provide a better view of Earth, at least for now.

"Goodbye from Apollo 8," he said, waving a hand.

And with that, the first broadcast from the first men on their way to the Moon went dark.

———

About ninety minutes after the telecast ended, Anders began to tire, yet he was still as wired as if the spacecraft was just lifting off. Needing to sleep, he requested permission to take a Seconal, which Houston approved. Floating in his hammock, he tried to will himself into oblivion, but his mind was a moving checklist, tuned to the vibration of the spacecraft and its systems. How could he make sure nothing went wrong if he allowed his mental blueprints to go dark? But the Seconal was working, oozing over his brain and melding all the sounds in the cabin—the instruments, the radio, his crewmates—into monotone.

Above, Lovell told Houston he was going to throw a switch and . . .

"Jim—not that one!" Anders cried, wide awake, thrusting up a hand and stopping the action.

It wasn't that Anders didn't trust Lovell. But Lovell had been a later addition to the crew (after replacing Collins) and hadn't had the opportunity to learn the command module the way he and Borman had. Long ago, Anders had determined not to tolerate much help on systems from Lovell, despite a deep respect for his crewmates' competence and capabilities.

Anders finally drifted off, for perhaps an hour, then shook off the effects of the sleeping pill, climbed back into his seat, and started working again. By now, Houston understood that the sleep schedules they'd engineered into the flight plan had long since drifted away. From this point forward, the crew would sleep according to their needs.

The astronauts spent the next several hours cruising, checking their systems and navigation, and looking back at an ever-shrinking Earth. At around forty hours into the flight, CapCom Jerry Carr radioed a news bulletin to Apollo 8. After 335 days of captivity, torture, and starvation, the crew of the American ship USS *Pueblo* had been released by their captor, the Communist government of North Korea.

In January 1968, the *Pueblo,* a U.S. Navy intelligence-gathering ship disguised to look like a fishing vessel, had deployed to waters just outside North Korean boundaries. She'd been sent on a covert mission to intercept military communications from that Communist regime, but just a few days into the operation, crews from several North Korean gunboats opened fire on the American ship, killing one crewman before boarding the *Pueblo* and taking its remaining crew, including the commander, prisoner.

President Johnson considered several hard-line responses but opted to try diplomacy first. His advisers fashioned fallback plans in case the United States needed to take military action. One of those plans, code-named Freedom Drop, called for the use of nuclear weapons to obliterate Communist troops that might storm into South Korea during an American attack.

Negotiations for the release of the crew stretched on for months. Held in miserable conditions, many of the Americans were interrogated, beaten, tortured, and threatened with execution. Commander Lloyd M. Bucher was put before a mock firing squad in order to obtain his confession, which he refused to make. He was then told his crew would be executed, one by one, until he acknowledged American guilt—which he finally did. After American officials signed a confession and apology, the North Koreans agreed to release the crew.

"The big news right now," Carr radioed to the spacecraft, ". . . is that all eighty-two crewmen of the *Pueblo* have been returned. They walked across the Bridge of Freedom Monday night."

"Wonderful!" Borman replied.

Anders had a different reaction. He was happy for the Americans who'd been released, but he couldn't help but compare the incident to the one his father had endured. Arthur Anders had defended the USS *Panay* from an unprovoked Japanese air attack in 1937, refusing to give up his ship, manning guns and returning fire even as he was gravely wounded. By contrast, the *Pueblo* hadn't even fought back. There were good reasons for that—the crew hadn't been trained well for combat, were not well armed, had been taken by surprise, and were outnumbered. But all that had been true of the *Panay*, too. It was hard for Anders not to wonder whether the crew of the *Pueblo* might have tried a little harder, as his father had.

Out his window, Anders looked toward Earth, now 165,000 miles away. From here, it was hard to pick out North Korea, or South Korea, or any countries at all.

The astronauts continued to sleep in fits and starts as the flight neared its two-day mark. In Houston, the wives maintained their squawk box vigils, listening for telltale signals in their husbands' voices—the subtle cues they first learned to hear when the men were teenagers—that would reveal how they really felt. So far, everyone seemed to sound good, though Susan, Valerie, and Marilyn each wondered if her husband was getting enough to eat.

At nearly forty-seven hours into the flight, Lovell provided a status report to Mission Control. Each of the men today had ingested between 40 and 60 ounces, or "clicks," of water (so called for the squirt gun contraption that dispensed it), along with some rehydrated and solid foods. By now, the crew had discarded NASA's feeding plan as completely as it had the sleeping plan. They were supposed to eat four meals a day, but it was clear that Lovell's appetite was biggest and that each man preferred some foods to others. The crew took to swapping—Anders would trade almost anything for apricot cubes, Lovell for bacon squares. No one could give away his beef and egg bites, which left a pasty coating on the tongue. Much of the food had to be reconstituted, either by injecting water into pouches or by mushing it with saliva in one's mouth.

At Mission Control, the doctors were not yet convinced that Borman, or even his crewmates, were operating at full strength. By their estimates, the astronauts still hadn't consumed enough food or water, or slept enough hours. But what was NASA to do? They were dealing with three grown men, each of whom was risking his life for his country, who now didn't want to eat their beef and egg bites. If the men began to starve, they'd eat.

Forty-eight hours into the flight, Apollo 8 was two-thirds of the way to the Moon. By then, it had slowed to a velocity of just over 2,400 miles per hour, about 10 percent as fast as when it departed its parking orbit around Earth. That kind of reduction frustrated thirteen-year-old Jay Lovell, who'd thought it just about right when the spacecraft had been going more than 24,000 miles per hour when it left Earth orbit.

It was now just past sunrise on Monday morning in Houston, and Valerie Anders was awake and feeding her five children. "My dad will be one of the first men on the Moon," six-year-old Greg told her. Through her front window, Valerie could see the mass of reporters on the lawn, already gathered in the near-freezing December air. She made them a pot of coffee, which she put in her garage along with a stack of cups. And she sent eleven-year-old Alan to rake leaves in the yard.

At the Borman home, Susan still couldn't eat. Her sons began to worry. "You've gotta have something," they said, but Susan couldn't stomach it. Fred took up a forkful of potato salad delivered by some good soul.

"Open up for the airplane!" he said, making the food swoop and loop with engine sounds, which was just what Susan had done for him when he was a little boy. If the method had been good enough for Fred and Ed, now it had to be good enough for their mother. She laughed, opened up, and took a bite.

That morning, the Borman boys threw on some camouflage gear, grabbed their shotguns, and left the house to go duck hunting. The embedded *Life* magazine photographer sensed a great shot and asked Susan to pose with the boys before they hit the road. She did, reaching somewhere for the smile that had earned her an offer from the Ford modeling agency when she was in college, and finding it, if just for the moment.

Frank Borman, age 27, already an instructor at the fighter weapons school, Luke Air Force Base, Arizona, 1955. It was here that Borman taught elite young Air Force pilots to fly for America and defend her greatness. *Courtesy of Frank Borman*

Susan Borman, age 19, just before her marriage to Frank in 1950. *This girl,* Frank thought, *can handle anything. Courtesy of Frank Borman*

Marilyn and Jim Lovell, aboard the U.S. Naval Academy schooner *Freedom*, just after their wedding in 1952. "I don't know how to dance," Marilyn told Jim when he first asked her out. "I don't either," he replied. "We'll learn together." *Courtesy of Jim and Marilyn Lovell*

Bill and Valerie Anders on a ranch in Colorado, around the time of Apollo 8. Bill's mother had wanted him to date an admiral's daughter, but Bill just wanted to be with Valerie.
Courtesy of Bill and Valerie Anders

Three of NASA's titans. Left to right: Robert Gilruth, George Low, and Chris Kraft. It was Low who master-minded the daring change in mission for Apollo 8; Kraft and Gilruth risked everything to support his idea. "It took more courage to make the decision to do Apollo 8 than anything we ever did in the space program," Kraft would say decades later. NASA

НАШ ТРИУМФ В КОСМОСЕ ГИМН СТРАНЕ СОВЕТОВ!

"OUR TRIUMPH IN SPACE IS THE HYMN TO THE SOVIET COUNTRY!"
The race to the Moon was an existential battle years in the making. Even on December 6, 1968, just days before the Soviet and American launch windows opened, the contest between the two superpowers was still too close to call. (Soviet propaganda poster, early 1960s.)

Apollo missions required a year or more of planning and training. Racing to preserve President Kennedy's deadline—and to beat the Soviets to the Moon—NASA had just four months to make Apollo 8 work. Here, the astronauts map out their lunar course, with Lovell pointing to a mountain he planned to name for his wife. NASA

The crew of Apollo 8—Anders, Borman, and Lovell—during water egress training in the Gulf of Mexico. Borman believed he had the best crew ever assembled at NASA. *NASA*

Apollo 8 crew taking a break during simulator training. *NASA*

Lovell, after an altitude chamber test at the Cape. Since boyhood, he'd dreamed of exploring and pioneering, and there seemed to him no better way to do it than by becoming the first man ever to fly to the Moon. *NASA*

The Apollo 8 spacecraft and Saturn V rocket being moved by NASA's Crawler-Transporter to Pad 39A at the Kennedy Space Center. Until Apollo 8, the Saturn V had flown only twice, both times unmanned. Its first test had been a success, its second a near disaster. *NASA*

Susan and Frank Borman in the living room of their home just before the launch of Apollo 8. Susan believed, with one hundred percent certainty, that her husband was going to die on the mission.
Ralph Morse/Getty Images

Jim Lovell kisses son Jeffrey at a beach near the launch pad while wife Marilyn and eldest son Jay look on. In a few hours, Apollo 8 would launch for the Moon. *Yale Joel/Getty Images*

Breakfast on the morning of launch. Clockwise from far left: Borman, Lovell, Anders, security chief Charles Buckley, head of mission support Hal Collins, Buzz Aldrin (red shirt), George Low. *NASA*

Borman suiting up for the launch of Apollo 8. *NASA*

The Apollo 8 crew, led by Borman and followed by Lovell and Anders, depart the Kennedy Space Center before dawn en route to the launch pad on the morning of December 21, 1968. *NASA*

The launch of Apollo 8, mankind's first journey to the Moon. *NASA*

Marilyn Lovell watches the launch of Apollo 8 with three of her children (from left, Susan, Jeffrey, and Barbara). Marilyn wanted to be as close to it and as much a part of it as possible. *NASA*

Borman aboard Apollo 8 as the spacecraft streaks toward the Moon. The command module measured just thirteen feet by eleven feet and was filled with equipment. *NASA*

Alan Anders, age 11, watching coverage of his dad's mission on TV.
Courtesy of Bill and Valerie Anders

Mission Control on day three of Apollo 8's journey to the Moon. The monitor shows an image that people throughout the world could see on their own television sets: Earth from a distance of 176,000 miles. For the first time in history, mankind was looking back at itself. *NASA*

The view through Anders's camera. Until Apollo 8, no human had ever laid eyes on the far side of the Moon. *NASA*

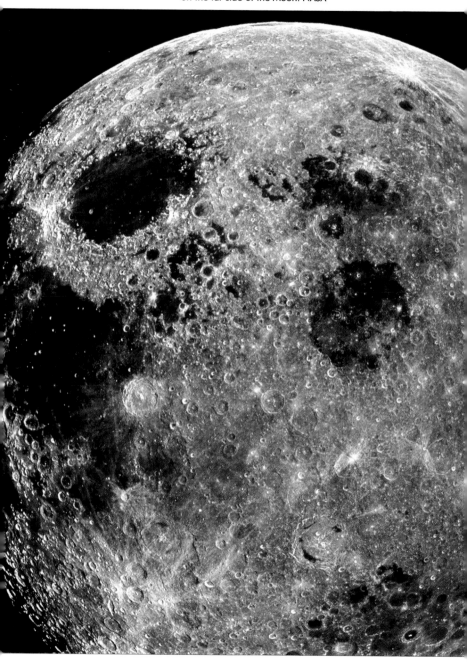

Susan Borman watching television coverage of Apollo 8, which was scheduled to orbit the Moon during Christmas. She composed Frank's eulogy around the time this photograph was taken because she needed to be ready. *Lynn Pelham/Getty Images*

Earthrise, as photographed by Bill Anders on Christmas Eve, 1968. Fifty years later, this photograph remains among the most influential and impactful of all time. To Anders, it seemed strange—the astronauts had come all this way to discover the Moon, and yet here, they had discovered the Earth. *NASA*

To many at NASA, leaving lunar orbit was the single most dangerous aspect of the Apollo 8 mission. Here, Valerie Anders and Susan Borman react at the Borman home as Jim Lovell confirms the crew has left the Moon: "Please be informed," Lovell radioed to Earth, "there is a Santa Claus." (Note the squawk box by Susan's arm). *Lynn Pelham/Getty Images*

President Johnson, like millions around the world, watches Apollo 8's reentry and splashdown live on television. *Bettmann/Getty Images*

Borman thanks sailors aboard the USS *Yorktown* after splashdown, while Anders and Lovell look on. Note that Borman is freshly shaved; before the mission, he'd asked that an electric razor be available aboard the recovery helicopter that would pluck the astronauts from the ocean. *NASA*

Anders and his family greet well-wishers at Ellington Air Force Base near Houston upon returning from their lunar mission. More than two thousand people showed up at 2:00 A.M. to welcome the astronauts home. *NASA*

Bill Anders, just after the return of Apollo 8. At the time he was likely the most famous photographer in the world. *Yousuf Karsh*

The crew of Apollo 8 (left to right: Bill Anders, Jim Lovell, Frank Borman) with their spacecraft at Chicago's Museum of Science and Industry, April 2018. The astronauts had come to help launch this book and celebrate the 50th anniversary of their historic mission. (Video of the event can be seen at robertkurson.com/rocketmen.) *JB Spector/Museum of Science and Industry, Chicago*

The boys sneaked out through the back fence, where they rendezvoused with Fred's car, which he'd parked near a neighbor's house. They intended to go to a friend's farm in the country, but by that time the press had gotten wise to the teenage Houdinis and gave chase. Fred mashed the gas pedal to the floor, making several tight turns and leaving the reporters in the dust. Two days earlier, Frank Borman had become the fastest man in history. Yet the boys knew that their father would have reached an arm down from space and strangled them if he'd known they'd been speeding.

In Timber Cove, Marilyn Lovell's concern had shifted to her son, Jay, who had begun to complain of stomach pains. Spiriting him to a neighbor's car and hiding him under blankets to avoid being trailed by the media, she drove to NASA, where doctors attributed his symptoms to the excitement that comes from having a father on his way to the Moon. While Marilyn and Jay were gone, two-year-old Jeffrey did his best to stand in for his mother, opening the door and answering questions from reporters while wearing a plaid jumper and his toy astronaut helmet, occasionally looking skyward in case his dad flew by. Now 186,000 miles above him, Apollo 8 was at precisely the distance at which it took light (and radio transmissions) one full second to reach Earth. If Jeffrey had seen his father flying by, the image would have come from history, not the present.

By the looks of things, Apollo 8 was in cruise mode, so Valerie Anders decided to visit Mission Control—for a change of scenery, and to feel closer to Bill. There she took George Low's hand and told him how grateful she was that Bill and his crewmates were being looked after so well. Low's blend of intellect and calm had made him a favorite of Valerie's, and his was the perfect hand to hold while she watched the green blip on the distant screen inch closer to the Moon.

A few minutes later, she heard Bill announce that he was going to "take a little snooze here for a while" and then sign off. She smiled and whispered something to a NASA official, who walked over to the public affairs officer, who in turn walked over to Collins. Shortly after that, Collins radioed to Borman aboard Apollo 8.

"Paul tells me Valerie is over here and wishes Bill a happy nap."

"Okay, thank you," Borman answered. "Tell her that he makes us tired sometimes, too, will you?"

Collins laughed.

"Roger. I will deliver a modified version of the message."

At 53 hours into the flight, the guidance and trajectory specialists in Houston were delighted by the precision of the journey. After Apollo 8 left Earth orbit, Houston had planned for up to four midcourse corrections for the coast to the Moon. But the first, accomplished during the brief test of the SPS engine about eleven hours into the flight, had been so accurate that the next two were dispensed with. Now, as they neared the Moon, only one tiny adjustment would be required, and that would use the spacecraft's small control thrusters.

The second live television broadcast was scheduled to begin in an hour. Despite the best advice from experts—on filters, lenses, switches, brackets, interior lighting, and exposure levels—Borman remained skeptical.

"I bet the TV doesn't work," he told Collins.

Just before 3:00 P.M. EST, with Apollo 8 at an altitude of 200,000 miles, the crew got ready to transmit. A few seconds later, a rounded edge appeared on television screens across America, then disappeared. Collins radioed instructions to the spacecraft, but the screen stayed gray.

Suddenly, an orb drifted dead center into the middle of the picture, and the shape of clouds and continents sharpened into view. For the first time in history, mankind was looking back at itself—at all of itself. Every human culture and language and idea and conflict and difference fit into a single picture.

In Mission Control, in living rooms, in hotels and bars and bus shelters across America, in the astronauts' homes, people fell silent.

Then Lovell spoke.

"Houston, what you are seeing is the Western Hemisphere. Looking at the top is the North Pole. In the center—just lower to the center is South America—all the way down to Cape Horn. I can see Baja California, and the southwestern part of the United States. There's a big, long cloud bank going northeast, covers a lot of the Gulf of Mexico, going up to the eastern part of the United States, and it appears now that the East Coast is cloudy. I can see clouds over parts of Mexico; the parts of Central America are clear. And we can also see the white, bright spot of the subsolar point on the light side of the Earth."

The broadcast was in black-and-white, so Collins asked about the colors.

"For colors, the waters are all a sort of a royal blue," Lovell said. "Clouds, of course, are bright white. The reflection off the Earth is much greater than the Moon. The land areas are generally a brownish, a sort of a dark brownish to light brown texture."

Watching out the window, Anders almost forgot the camera in his hand. To him, Earth seemed almost to transcend reality, its color and brightness and clarity beyond what one could see when one was actually on the planet. He could hardly imagine a more beautiful image.

After a few more minutes, Lovell added to his description.

"Mike, what I keep imagining is, if I'm some lonely traveler from another planet, what I think about the Earth at this altitude, whether I think it'd be inhabited or not."

"Don't see anybody waving, is that what you're saying?" Collins replied.

"I was just kind of curious whether I would land on the blue or the brown part of the Earth," Lovell said.

"You better hope we land on the blue part," Anders chimed in, causing an eruption of laughter at Mission Control.

Near the end of the twenty-two-minute telecast, Collins asked Borman a question millions of people had on their minds.

"How about the Moon, Frank? Is it visible through one of your other windows? Could you get it visible with a small maneuver?"

"Negative," Borman answered. "I think we'll have to save the Moon for another time."

A few minutes later, the show came to a close. So compelling was the view of Earth that the crew hadn't once thought to turn the camera on themselves.

"Okay, Earth," Borman said. "This is Apollo 8 signing off for today."

And with that, the astronauts disappeared back into the sky.

The astronauts' younger children dispersed when the broadcast ended, running outside to play, fixing sandwiches in the kitchen, pulling toys from their toy chests. Hearing the clamor of the reporters on their lawns, Susan and Valerie went outside their homes to answer questions from the press. Only Marilyn didn't move from her spot in front of the television. She just stared at the screen, trying to process the distance between her and her husband.

Focus at Mission Control turned to the giant projection screens at the front of the room. By the estimates of trajectory and orbital mechanics specialists, Apollo 8 was about to cross what some called the equigravisphere, the point at which Earth and Moon exert an equal pull of gravity. It had taken until now, about five-sixths of the way to the Moon, to reach the equigravisphere, a testament to the dominance of Earth in its finely balanced relationship with its smaller satellite. There would be nothing to mark the place in space, no bump or jolt to the spacecraft. But in its silence, the crossing would make a thundering announcement—for the first time, man had become captured by the pull of another celestial body.

The men in Mission Control had bet on the event; the winner would be the one who most accurately predicted the moment of crossing. It wasn't an easy guess, as the line changed depending on the distance between Earth and Moon at the moment, the phases of the Moon, and other factors. But controllers would know it when the moment arrived, because for the first time in more than two days, the spacecraft would stop slowing down and begin to gain speed.

At around this time, Mission Control received a visit from Marilyn Lovell, who'd hitched a ride with a NASA representative. In the viewing area, Robert Gilruth, the director of NASA's Manned Spacecraft Center, greeted her and sat down to talk. Perhaps he sensed her apprehension, or maybe he was just being friendly, but he did not rush off to attend to his pressing duties, even at this historic moment.

At 55 hours, 38 minutes into the flight, all eyes in Mission Control, including Marilyn's, turned to the big screens. Controllers checked the numbers—Apollo 8 was 202,700 miles from Earth, but its speed, about 2,200 miles per hour, was no longer dropping. Moments later, a light flashed on the screen.

"My God," one of the young computer specialists said. "Do you know what we just saw?"

And the room, transfixed, did know. The light meant that Apollo 8 was no longer a part of this world.

Chapter Seventeen

●

RACING THE MOON

SCREENS AND CHARTS AT MISSION CONTROL CHANGED. NO longer was the spacecraft 202,700 miles from Earth; it was now 38,760 miles from the Moon. And it was picking up speed, passing 2,700 miles per hour and gaining by the minute.

At their consoles, controllers made printouts of their displays to commemorate the moment. Someday they would show these papers to their grandchildren and tell them what they'd seen.

A few minutes later, Collins radioed Apollo 8 with an update on their recent television broadcast.

"We are having a playback of your TV shows and are all enjoying it down here. It was better than yesterday because it didn't preempt the football game."

"Don't tell me they cut off a football game," Borman said. "Didn't they learn from *Heidi*?"

Just a month earlier, as millions of Americans watched the

New York Jets and Oakland Raiders battle into the final minute of a spectacular game, NBC stuck to its strict broadcast schedule, switching over at 7:00 P.M. to *Heidi,* a film about a young girl who was living with her grandfather in the Swiss Alps. Viewers erupted in protest, flooding the network with irate calls (and threats) and blowing out twenty-six fuses on the NBC switchboard while the Raiders scored two touchdowns in nine seconds to pull off a miracle come-from-behind win. The next day, *The New York Times* ran a front-page story on the debacle, and David Brinkley addressed it on the evening news—then showed tape of the game's last minute. Even on this pioneering trip to the Moon, Borman wanted nothing to do with messing with his beloved game of football.

Thirty minutes after she'd arrived, Marilyn left Mission Control with astronaut John Young, who was going to drive her home.

"Have you seen Susan yet?" Young asked, then offered to take her over to the Borman house.

When they arrived, Marilyn found a familiar scene—loads of visitors, trays of food, kids pinballing between rooms, squawk boxes chirping. The only thing missing was Susan.

"She's in the bedroom," someone said. "I'll tell her you're here."

Marilyn sat in the living room and waited, chatting with other visitors and fixing herself a drink. She kept waiting, for thirty minutes, an hour. After two hours, Susan still had not emerged.

I'm part of this just as much as you are, Marilyn thought. *My husband is on this flight, too.*

Marilyn didn't know how painful things had been for Susan after the Apollo 1 tragedy. She didn't know how clearly Susan pictured herself as a widow in the coming hours. If Marilyn had known any of this, she would have understood and would have tried to help. But Susan never showed that vulnerability to anyone, not even to Frank. As Marilyn waited, Susan remained curled up on the bed in her bedroom, listening for her husband's voice on the squawk box. When Marilyn left, she left with hurt feelings.

Back at home, Marilyn found her house oddly empty. She poured herself a scotch on the rocks, sat on a stool at the wood-paneled bar in the family room, and sobbed. In just ten hours, Apollo 8 would disappear behind the Moon. How had she been so confident all this time? Her husband was disappearing *behind the Moon*. And that meant he might never come home.

At almost exactly two and a half days into the flight, Apollo 8 prepared for just its second—and final—midcourse correction burn of the outbound leg. It would be accomplished by firing four thrusters on the spacecraft, each of which could produce 100 pounds of thrust. That was only the tiniest fraction of the force that had been required to get Apollo 8 off the launchpad, but it was all the vehicle would need for eleven seconds as it refined its line to the Moon.

"Okay, stand by," Borman called to his crewmates.

"Burn," Lovell said.

"Burning," Anders confirmed.

Eleven seconds later, it was done. Houston analyzed the telemetry—the correction had been nearly perfect, and it was just a matter of riding the ship for another eight hours until lunar rendezvous. Despite such close proximity, the crew still could not see its target. To all of them, it felt like sitting with their backs to the screen in a movie theater during a terrific thriller.

In Houston, the wives began to prepare for when the spacecraft reached the Moon, scheduled for 4:00 A.M. Houston time, when Apollo 8 would attempt a complex maneuver known as Lunar Orbit Insertion, or LOI. Engineers, mathematicians, physicists, and other scientists had spent years developing the calculations and determining how to make the maneuver work. But on its face, LOI was easy to understand.

At 69 hours into the flight, Apollo 8 would pass just in front of the Moon, missing its surface by only 69 miles. That altitude had been chosen for a reason. On future landing missions, it would be close enough so that the lunar module shuttling astronauts to the lunar surface and back wouldn't require a massive amount of propellant, but far enough away to make it unlikely that the spacecraft waiting in orbit above would crash into the Moon.

If Apollo 8 did not fire its SPS engine—or if the engine failed to ignite—after passing behind the Moon, lunar gravity would cause it to slingshot around the far side and head back to Earth, requiring only minor course adjustments in order to hit its reentry corridor and splash down in the Pacific Ocean. NASA had chosen this free return, figure eight trajectory in case of engine failure or other in-flight problems.

But NASA planned for Apollo 8 to orbit the Moon. To enter lunar orbit, the spacecraft had to slow itself down enough to be captured by the Moon's gravity. The only way to do that was to fire the SPS engine against the direction of its travel, for just the right amount of time—about four minutes—and with just the right amount of thrust.

If the engine fired for too short a period, or without enough thrust, the spacecraft might still slingshot around the Moon but emerge on an improper trajectory, one that might cause it to burn up on reentry into the atmosphere or miss Earth entirely. Or it might be cast out into space without enough power or propellant to reverse course and come back. Or it might enter a lunar orbit off-kilter enough to cause it to crash into the Moon.

If the engine fired for too long, or with too much thrust, the results might be even worse. That would slow down the spacecraft so much that the Moon's gravity would overcome the ship and cause it to plummet into the lunar surface.

The SPS was the same engine that had failed to build up proper thrust on its test fire eleven hours into the flight, owing to the suspected helium bubble in a propellant line. Kraft and the controllers in Houston believed the problem had worked itself out, but that was just a best guess. They wouldn't know for sure until the astronauts tried to light the engine again behind the Moon.

No aspect of the Lunar Orbit Insertion maneuver was easy. In training, the crew and controllers crashed into the Moon time and time again. Sometimes controllers became so anxious that they aborted prematurely or took needless emergency action, fracturing their confidence and planting doubts about their ability to work as a single cohesive unit. So shaken did some controllers become that they sometimes denied they'd made a mistake. To Kraft, that was even worse than the error. When denials happened, he'd stop the session, take the controller aside, and say, "I don't want any bullshit from you anymore. If you make a mistake, you say you made a mistake." Kraft demanded truth, and he demanded that everyone perform the maneuver again and again and again. With a few hours to go, he believed his team was ready.

Lunar Orbit Insertion involved an emotional component, too. After flying past the leading hemisphere of the onrushing Moon, the spacecraft would disappear behind its far side and lose contact with Earth, as all signals to and from the ship would be blocked by the Moon. For about thirty-five minutes, no one at Mission Control would have any idea whether the SPS engine had fired or performed well; no one would be able to monitor the spacecraft or its systems, no one would be able to talk to the astronauts. Controllers could only watch their clocks and hope Apollo 8 emerged from the far side exactly when it was supposed to. If it came out any earlier or later than that, something had gone wrong.

Anders had imagined he would watch the Moon closing in, as the pilots did in the film *2001: A Space Odyssey*, until it filled the sky. Instead, he saw emptiness. Even Lovell, with the wide field of view from his telescope and sextant, couldn't catch a glimpse of the Moon. Due to the position of the Sun (and the glare it caused) and the position of the spacecraft's windows (facing mostly toward black space), it had been all but impossible for the crew to spot its target.

"As a matter of interest, we have as yet to see the Moon," Lovell radioed to Houston.

"What else are you seeing?" asked CapCom Jerry Carr.

"Nothing," Anders replied. "It's like being on the inside of a submarine."

In Houston, in the middle of the night, the astronauts' wives turned their squawk boxes just loud enough to hear without awakening friends and family who were curled up on couches and in chairs throughout their homes. One by one, these supporters awoke to help the wives through Lunar Orbit Insertion. In the fog of nerves and excitement, few realized that it was now officially Christmas Eve.

Despite the hour, approaching four in the morning in Houston, visitors began to crowd into the viewing room at Mission Control. A hundred people sardined themselves into this room designed for far fewer, but all respected the flashing sign that requested QUIET PLEASE as Lunar Orbit Insertion drew near.

It was time for Mission Control to make a final decision. One by one, Flight Director Glynn Lunney polled each of his controllers, looking for a simple Go or No Go. One by one, they gave him their answer. Lunney looked at Carr, who radioed to the spacecraft, now just over three thousand miles from the Moon.

"Apollo 8, this is Houston. At 68:04, you're Go for LOI."

"Okay," Borman answered. "Apollo 8 is Go."

"You are riding the best bird we can find," Carr said.

Thirty minutes remained until Lunar Orbit Insertion. Controllers continued to make final checks of the spacecraft and its systems, and to grow more nervous by the minute.

George Mueller had thought 69 miles was cutting it far too close to approach or orbit the Moon. "You don't know that you're that accurate," Mueller had told Kraft when the mission was planned. "You don't know that you can hit the Moon within sixty-nine miles as you're aiming at this thing two hundred and forty thousand miles away. You don't know that your radar is that good. You don't know that your tracking is that good." Kraft agreed that orbiting at a higher altitude would decrease the chance of error and catastrophe. But that wouldn't have allowed NASA to best prepare for a lunar landing. So 69 miles it would be.

Mueller wasn't the only one worried. Pacing the back row at Mission Control, the lead flight director, Cliff Charlesworth, who was off duty at the time, kept thinking, *I know all our guidance systems are accurate, and we tracked it properly, and all the mathematicians in the world have looked at this thing. But sixty-nine miles is pretty close . . .*

In a back room, John Mayer, chief of the Mission Planning and Analysis Division, began to receive visitors—Bob Gilruth, George Low, and other top managers, who'd arrived with a pressing, semiserious question:

"How sure are you we're going to miss the Moon?"

On board Apollo 8, the crew had the same concern. Their spacecraft was now traveling more than 5,000 miles per hour. For its part, the Moon, 2,160 miles in diameter, was moving at more than 2,000 miles per hour. Could anyone really guarantee the ship wasn't going to end up smashing into the massive orb?

At one console, Flight Dynamics Officer Ed Pavelka calculated the SPS burn data Apollo 8 required for its Lunar Orbit Insertion. Nearby, Jerry Bostick, chief of the flight dynamics program, the team responsible for the trajectory and guidance of the spacecraft, watched him check and recheck his calculations, not once or twice but nine or ten times, before passing them along to Carr for transmission to the astronauts.

Five minutes remained until Apollo 8 met the Moon. At home, Susan, Marilyn, and Valerie hung on every word from the squawk box.

"Apollo 8, Houston. Five minutes . . . all systems Go. Over," Carr radioed to the crew.

"Thank you. Houston, Apollo 8," Borman replied.

"Roger, Frank," Carr said. "The custard is in the oven at three fifty. Over."

That was a secret message from Susan to Frank. Long ago, he'd told her, "You worry about the custard and I'll worry about the flying"—separating their duties was the only way to survive the toll a test pilot's career exacted from a marriage and family. She'd wanted to let him know that all was good at home at a time he might need to hear it most.

"No comprendo," Borman told Carr.

Susan couldn't tell whether Frank hadn't understood the words or had forgotten the reference. All she knew for sure was that she couldn't reach him.

Two minutes remained until the spacecraft, now moving at 5,125 miles per hour, went behind the Moon. Since lift-off, Apollo 8 had traveled 240,000 miles, and the Moon had traveled 150,000 miles, to make this rendezvous.

"One minute to LOS [loss of signal]," Carr radioed to Apollo 8. "All systems Go."

"We'll see you on the other side," Lovell said.

Outside Anders's window, any trace of sunlight had disappeared, and as his eyes adapted to the intense darkness he began to see stars, it seemed like a million of them, so many he couldn't even pick out constellations. The sight took his breath away. He looked to his right, through the window beside him, hungry for more, but suddenly there were no stars anymore—all of them had gone dark. There was just a giant black hole, as if part of the universe had vanished. The hair on the back of Anders's neck stood up, and for a moment it felt as if his heart had stopped, until he realized that he wasn't looking at a missing piece of the universe at all.

He was looking at the Moon.

A few seconds after that, Apollo 8 disappeared behind it.

Chapter Eighteen

●

OUR MOST ANCIENT COMPANION

THE MOON IS APPROXIMATELY 4.5 BILLION YEARS OLD, ABOUT the same age as Earth. The Moon is not a planet but a satellite, and a unique one in the solar system, much larger than other satellites that orbit solid, rocky planets (usually such giant moons revolve around gaseous bodies).

While the Moon is one-quarter the size of Earth, its mass is only about 1 percent of that of Earth. Gravity on the Moon acts with just one-sixth the strength that it does on Earth. Every year, the Moon drifts about an inch and a half farther from Earth as a result of the acceleration effects of Earth's ocean tides. As a result, the rotation rate of Earth is gradually reducing.

Unlike Earth, which currently rotates every twenty-four hours, the Moon rotates around its axis just once a month. Because it also circles Earth once a month, the Moon's near side is always the side facing Earth. It wasn't until 1959, when an unmanned Soviet spacecraft snapped grainy photographs of the far side, that anyone had any idea what it looked like. If all went according to plan, the crew of Apollo 8 would become the first humans ever to lay eyes on the far side of the Moon.

The far side is often referred to as the dark side, but that is a misnomer. All sides of the Moon receive sunlight and experience days and nights. The Moon's slow rotation on its axis does mean that areas can stay in sunlight, or in darkness, for nearly two weeks.

In many ways, the story of the Moon is a story of its two sides. The near side, which has been facing Earth for billions of years, is marked by dramatic contrast between light and dark sections, which can easily be seen from Earth with the naked eye. The light areas are the highlands, covered with craters and rolling with hills and mountains that can rise miles above the surface. The dark areas are called mare (*mare* is Latin for "sea"; the plural is *maria*). The maria cover about one-third of the near side and are much smoother than the highlands, with far fewer craters, an unusual coda to a violent story.

For billions of years, the Moon was bombarded by asteroids, comets, and other debris. The lunar surface is a record of those impacts, each of its round craters, with its sharp rim and rising edges, a snapshot of a collision. Some of the most massive impacts excavated huge basins with fractured floors that allowed molten magma to ooze up from the Moon's mantle and fill them in, solidifying into a dark, smooth basalt that became the maria. These giant impacts were the climax of an era of bombardment that marked the Moon's initial evolution. The formation of the maria occurred long after the basins were formed, in some cases perhaps by a half billion years or more, so they are younger than the light-colored highlands. The maria are also relatively free of craters, suggesting that the lunar landscape appears much the same today as it did three billion years ago.

The far side of the Moon is very different from the near side. It has almost no maria—just 1 percent of its surface area is dark "sea" compared to 30 percent on the near side. Instead, almost the entire far side is covered in heavily cratered, light-colored highlands. Many scientists think this is the result of a lesser concentration on the far side of radioactive elements, which produced the volcanic activity that created the maria on the near side. Experts also believe that the far side has a thicker crust, which would have made it more difficult for even the largest asteroids and comets to break through to reach the magma below.

The far side is also more mountainous, with its highlands rising higher above the surface than on the near side. And it is more heavily cratered because it has so few maria, the "seas" of hardened magma that covered over so many impacts on the near side.

Once the differences between the near and far sides are accounted for, it becomes easy to describe the Moon as a whole. Perhaps the Moon's most distinctive feature is its craters. These range in size from microscopic to the South Pole–Aitken basin, which measures 1,550 miles across and 5 miles deep. There are possibly more than a million craters at least a half mile wide on the Moon. Even at close range, and with optical instruments, the crew of Apollo 8 would never be able to count all the craters on the Moon.

The current impact rate from meteoroids and other debris is just about one ten-thousandth of what it was during the late heavy bombardment period about four billion years ago, when the basins formed. Today, that equates to about a hundred impacts a year by objects weighing between a fraction of a pound and a ton.

As a result of the constant bombardment of asteroids and comets, the vast majority of the lunar surface is coated in a mixture of powdery dust and pulverized rock fragments known as *regolith*. This top layer might be as shallow as six feet at the maria, or as deep as thirty feet in the highlands. For years, NASA planners worried about whether a spacecraft, or even an astronaut, might sink beneath the regolith and disappear. In the mid-to late 1960s, unmanned probes sent by NASA answered that question: The regolith was sturdy enough to support lunar landings, even if spacecraft would settle into it a bit and men might make footprints with their boots.

The Moon's crust—its rocky, rigid outer layer—is much thicker (35–60 miles) than Earth's (3–20 miles), remarkable given the relative sizes of the two bodies. The opposite is true of the Moon's core, which is much smaller and lighter (3 percent of total mass) than Earth's (one-third of total mass).

The Moon isn't a perfect sphere. It's difficult to see from Earth, but the Moon is a bit squashed at the poles, with a slight bulge at the equator, which points toward Earth. That bulge is evidence of Earth's grip on the Moon. The Moon's gravitational pull on Earth is equally important; without it, Earth would wobble on its axis and lose its moderate climate. Summer temperatures could exceed 200 degrees Fahrenheit. Much of Earth could sink beneath water. Spinning faster without the Moon's grip, Earth days might last just eight hours, winds would reach hurricane strengths, and life would be difficult, if not impossible.

There is essentially no atmosphere on the Moon; its gravity isn't strong enough to keep hold of an envelope of gases. Without an atmosphere, the Moon cannot trap or filter heat. On the side facing the Sun, temperatures can rise to 240 degrees Fahrenheit; on the other side, they can plummet to minus 290 degrees Fahrenheit. The lunar surface and surroundings are in a vacuum, which should make the Moon absolutely dry and devoid of water. Yet recent probes proved that there is water ice in the regolith of craters at the lunar south pole, which exists in eternal shadow. If humans someday set up a colony on the Moon, they'd probably start at these craters, where water is most likely to be.

On Earth, there is little sign of the bombardment the planet has received from meteorites and other space debris. Rain, wind, plate tectonics, glaciers—all of these factors have worked over eons to erode or bury the evidence of these impacts on Earth. On the Moon, nothing is churned or worn away. The scars from objects that strike the Moon are preserved; this is true even for objects that arrived during the earliest days of the solar system. Examining particles blown onto the Moon by the solar wind might reveal much about the young Sun, when that star was just born. And examining particles thrown off by Earth onto the Moon would tell us about our own history—and ourselves.

Little is known about Earth's first billion years, the time when primitive life originated on the planet. Earth meteorites preserved on the Moon could provide a window back to that time, giving us a glimpse of the ages from which we came, the stuff from which we are made. But there would be no way to examine Earth meteorites embedded in the Moon without space travelers who could bring them back to us—without humans brave enough to climb into a spacecraft, light an engine with the power of a nuclear bomb below them, and land on our most ancient companion. And one must wonder if, in the future, a similar push by bold adventurers, this time beyond the Moon and into the universe, might bring back another kind of knowledge about ourselves, one that we might not yet have the capacity to imagine but that might transform us fundamentally.

Chapter Nineteen

●

EARTHRISE

FOR MONTHS, BORMAN HAD BEEN FIXATED ON A PARTICULAR moment in the flight plan: the instant when Apollo 8 would lose radio contact with Earth as it slipped behind the Moon. This would not be the first time a space mission lost contact with Earth. In fact, every Earth orbital flight (Mercury, Gemini, and Apollo 7, as well as the Soviet flights) had long periods when the spacecraft was out of touch with all the ground stations due to Earth's curvature. Since the planet was not covered with ground stations, the crews on those missions spent most of their time in radio silence. But that was far different from losing contact with the home planet because another world got in the way, which was just about to happen with Apollo 8. NASA had calculated, to the second, when it expected its communications with Apollo 8 to go dead. If the planners were correct, it meant the ship was on its proper trajectory and was where it should be.

If radio contact lasted too long, however, it likely meant

Apollo 8 had been traveling too fast and had arrived at its rendezvous point with the Moon before the Moon had a chance to get there and block the transmissions. If the arrival was just a little early, the spacecraft might still be whipped around the Moon by lunar gravity, but at a much higher orbit than desired. If the arrival was earlier than that, Apollo 8 might head off in a trajectory away from the Moon that it couldn't reverse for lack of sufficient onboard propellant.

If, on the other hand, radio contact ended prematurely, it likely meant Apollo 8 had taken too long to reach its rendezvous point with the Moon. If the lateness of arrival was slight, the spacecraft would zoom past the lunar surface at an altitude lower than NASA had planned or deemed safe for the mission. If it arrived much later, Apollo 8 would smash into the Moon.

So it was with great anticipation—and some dread—that the astronauts focused on the clock as the spacecraft flew backward, its cone-shaped nose and windows facing away from the direction of travel, into blackness. At the Anders residence, Valerie listened with friends, her living room dark except for the glow of a Christmas tree and a crackling fire. At her home, Susan Borman huddled in the breakfast nook and put her ear to the squawk box.

With just one second to go before predicted loss of signal, Apollo 8 was still in contact with Houston.

Borman's stomach tightened.

Lovell and Anders stared at the clock.

The view out the windows became even darker.

The astronauts' headsets went silent.

Borman looked at the clock.

"Jeez," he said.

Radio contact had been lost at precisely the second NASA had calculated.

Borman could hardly believe it. Anders joked, "Chris [Kraft] probably said, 'No matter what happens, turn it off.'"

Anders had seen how concerned—obsessed—Borman had

been about this moment during training. It took a second for Borman to realize Anders was kidding. After that, Borman couldn't stop smiling. Another critical hurdle in the Apollo 8 mission had been cleared.

In Houston, controllers looked at each other with a sense of wonder and relief, shaking their heads and then shaking hands. Orbital mechanics—the way the universe ordered and moved itself—worked. And man had figured it out to the split second.

The relief at Mission Control was short-lived. In ten minutes, Apollo 8 would fire its Service Propulsion System engine in order to slow itself enough to achieve lunar orbit. The SPS had to work perfectly. And everyone remembered how the engine had fallen short of optimal performance during its brief test firing on the way to the Moon.

Ordinarily, controllers in Houston could rely on their consoles and readouts to provide reassurance that all was well with the spacecraft. But that wasn't possible with Apollo 8 behind the Moon. No one on Earth would know how well the SPS engine had performed, or even if it had ignited, until Apollo 8 came around and reappeared on the near side of the Moon. If all went well, that would happen in thirty-six minutes.

"Okay, this is a good time to take a break," Flight Director Glynn Lunney said. He wanted everyone back in twenty minutes.

Glynn, you idiot! Jerry Bostick thought. *We've got Americans behind the Moon and you want to take a break?* But in a moment he realized that Lunney was right. There was nothing anyone could do to help Apollo 8 while it was behind the Moon. So it was a good time, indeed the only time, to visit the bathroom, grab a cup of coffee, and come back prepared. Headsets were removed and placed on consoles, and a pilgrimage made to the men's room (there were no women's restrooms at Mission Control in 1968 because there were no women). Standing in line, the twenty-eight-year-old Bostick had never felt more helpless. *Did we give them the right data?* he wondered. *Is everything okay?*

Engineer Aaron Cohen, who'd worked with Borman on the re-design of the command module, felt his body tense up. Dick Koos, one of the SimSups who'd constructed the nightmare scenarios to train the astronauts and controllers during simulations, felt faint and feared he might pass out.

On board the spacecraft, Anders had a realization: Given the ship's orientation, he had become the first man ever to reach the Moon, beating his crewmates by a few centimeters. And then it hit all of them.

They had reached the Moon.

Since humans first walked Earth, the Moon had been their siren, lighted their way in darkness, remained their companion in the night. It hung at an eternal distance, yet pulled on men and women as it pulled on the oceans, calling to a primal instinct—to journey beyond one's home and explore the unknown. But the Moon had always been too far, always beyond reach.

Today, Borman, Lovell, and Anders had changed that. Today, on December 24, 1968, when humankind opened their eyes, three of their own had arrived.

Before firing the SPS engine, the crew had to run through their checklists and position the spacecraft so that the burn would put them into a proper orbit. Even now, they were just 400 miles or so above the lunar surface, yet they couldn't see anything in the blackness because the light of the Sun and its reflected shine from Earth were blocked by the Moon.

A few minutes later, the spacecraft emerged into sunlight, at just the moment NASA planners had predicted. With less than two and a half minutes to go before SPS ignition, Lovell called out:

"Hey! I got the Moon! Right below us!"

Anders pushed for a closer look, but all he could see were streaks of oil rolling down his window. Then it hit him. Those streaks weren't oil. They were lunar mountains.

"Look at that!" Anders said. "See it? Fantastic!"

It was the first time human beings had laid eyes on the far side of the Moon.

Borman's commander instincts kicked in.

"All right, all right, come on. You're going to look at that for a long time." He needed to keep the mission focused. They had a rocket to fire soon, one that had to work.

"Twenty hours—is that it?" Anders asked, sounding as though he could look forever. Inside, he could only say to himself, *That's the* Moon*!*

Lovell prepared for the firing of the SPS engine, looking for an indicator from the display panel that signaled all was ready. Five seconds before ignition, he got it—the number 99 began flashing, the computer asking for the go-ahead to proceed.

Lovell pushed the button.

The astronauts felt a vibration, then the weight of their bodies pressed against their restraints as the spacecraft began to decelerate. The engine had lit, that much was certain. Now it had to burn against the direction of travel for just over four minutes to slow the ship's speed from around 5,100 miles per hour to less than 3,700 miles per hour, which would allow the Moon's gravity to capture the spacecraft for orbit. Inside the cabin, the men could hear the external thrusters firing as the computer worked to keep the craft straight.

Borman checked the instruments, which indicated the engine looked good. But no one on board seemed reassured.

"Jesus, four minutes?" Borman asked two minutes into the burn.

"Longest four minutes I ever spent," Lovell said.

The burn seemed never to end; the rocket just kept firing, the crew hyperaware of the fact that if it lasted even a little longer than necessary, it could smash the spacecraft and its crew into the Moon.

"Forty seconds left in the burn," Lovell called.

Anders picked up the countdown.

"Five . . . four . . . three . . . two . . . one . . ."

The computer was ready to shut down the engine.

Borman beat the machine to it.

"Shutdown," he announced, pushing the button. The spacecraft, and the men, settled back into weightlessness.

In Houston, the controllers weren't due back at their consoles for another seven minutes, but most of them had already returned and affixed their headsets. On board the spacecraft, the crew checked the delta-v—change in velocity—and could see they'd been captured by lunar gravity. Apollo 8 now belonged to the Moon.

Onboard readouts indicated that the spacecraft was in an elliptical orbit, ranging from a low point of 69.6 miles at its perigee to a high point of 195 miles at its apogee. Borman was astonished by the accuracy of the specialists who'd planned the flight. They'd predicted radio cutoff perfectly. And now they'd nailed the dimensions of the orbit to within a fraction of a mile.

Knowing their engine had made good, the astronauts were free to take a look out their windows. Below, they got their first clear view of the lunar surface.

At the sight, each man forgot his flight plan, even Borman. They leaned forward, pressing their faces against the spacecraft glass. To Lovell, the three of them looked like kids staring through a candy store window.

"It looks like a big—looks like a big beach down there," Anders said.

Despite his training in lunar geology, the far side of the Moon startled Anders. Long, oblique shadows showed the terrain to be much rougher than he expected, and with many more mountains, an impressive sight. He thought to himself that Stanley Kubrick hadn't gotten it right in his film *2001: A Space Odyssey*, in which he showed the Moon's surfaces to be sharp, angular, and scratchy. In real life, they looked sandblasted.

The size and number of craters was staggering. There were countless numbers of them, some as small as the eye could discern, others as wide as European countries. For years, scientists had argued about the cause of these impressions—volcanic activity or meteorite impacts? Most experts had come to the conclusion that craters were caused by meteorites. Anders scanned the surface of the far side but found no lava flows or other evidence of volcanic activity. He felt pleased to add his firsthand opinion to the debate: The craters had been made by meteorites, four billion years' worth, an endless bombardment from the solar system.

To Lovell, the surface looked like a concrete sidewalk that had been attacked by a man wielding a pickax, each wound rippling sand and particles around the impact point, so many craters they could never be counted. There was a harshness to the terrain, and no color, just grays and whites that went on forever. It wasn't beautiful, exactly, but to Lovell, the scene was awe-inspiring in its vastness and the story it told—a tale as old and as new as Earth and the Sun—and for that alone, it was beautiful to his eyes.

To Borman, spacecraft, rockets, and computers were the products of science, the logical advance of mankind. The lunar far side, however, seemed a dreamscape, straight out of science fiction. Nothing was lit like that on Earth, or even in one's imagination. Nothing was ever that alone. And yet he saw splendor in all of it, in the epochs of violence gone perfectly still.

The men could have watched the Moon for hours, but there was work to do. Borman would fly the ship, making certain the windows stayed in position for Lovell and Anders to perform their tasks. Lovell would take navigation sightings, confirm lunar landmarks, and assess potential landing sites on the near side for future missions. Anders would pull heavy photography duty while monitoring the spacecraft and its systems. Apollo 8 had ten revolutions to get all its work done, twenty hours total.

In Houston, the controllers were back at their desks, but they still didn't know that Apollo 8's SPS engine had performed well, or even whether it had fired at all. All they knew was that if it had failed to light, the spacecraft would appear just two minutes from now. For once, controllers rooted for their consoles to remain frozen; if any jumped to life now, it would mean Apollo 8 had come out too soon.

Kraft, Bostick, and others watched a clock that was counting down to the time the spacecraft would reappear if the burn had not taken place. It seemed antithetical at NASA to hope for nothing to happen when a countdown reached zero. But that was exactly the prayer in the church-quiet room.

When Apollo 8 failed to appear, waves of relief washed over Mission Control.

Now Houston had to jiu-jitsu its mindset. In ten minutes, Apollo 8 *had* to appear, right on time, or it likely meant disaster. A new countdown began, one that could be heard not just at Mission Control but also on the squawk boxes inside the homes of the three astronauts. While Marilyn Lovell remained surrounded by family and friends (and her priest), Susan Borman sat alone in her kitchen, lips pursed, trying to divine good or bad in the radio silence. Valerie Anders, teetering on the edge of sleep (it was not quite four thirty in the morning), believed the crew would appear right on time, a confidence that her nervous friends, who'd gathered to support her this predawn morning, must have appreciated.

As the countdown to predicted signal reacquisition reached one minute, Mission Control fell silent. CapCom Jerry Carr began to call to the spacecraft, broadcasting into a silent vacuum.

"Apollo 8, Houston. Over."

"Apollo 8, Houston. Over."

"Apollo 8, Houston. Over."

Finally, a voice came through the headsets at Mission Control.

"Go ahead, Houston, Apollo 8."

It was Lovell.

"Burn complete," he told his colleagues on Earth.

Mission Control exploded in cheers and applause. Apollo 8

had come around to the near side of the Moon. The contact had occurred within one second of NASA's estimate.

Chris Kraft's eyes began to mist over. He could see Bob Gilruth, the director of NASA's Manned Spacecraft Center, wiping away his own tears, hoping no one would see him cry. The two men embraced but couldn't speak; their throats were too swollen with emotion to talk.

Cheers also erupted in the astronauts' homes. Marilyn Lovell felt proud of her husband—his voice had been the first one broadcast from the Moon. To Valerie Anders, Lovell's simple statement—"Burn complete"—sounded like an ebullient "We're still here!" Susan Borman was happy that her sons were happy, but she felt no sense of relief. She'd seen this movie a thousand times in her head, and it always ended the same way.

Sixteen minutes after appearing on the lunar near side, Apollo 8 passed over the Sea of Fertility, an expanse roughly the size of France, visible with the naked eye to observers on Earth as one of the prominent dark patches on the Moon's eastern limb.

"What does the ol' Moon look like from sixty miles?" Carr asked the astronauts.

Lovell took the question. For the first time, man was about to hear man describe the Moon, not as a distant observer, but as an eyewitness.

"Okay, Houston," Lovell said. "The Moon is essentially gray, no color; looks like plaster of Paris, sort of a grayish beach sand. We can see quite a bit of detail. The Sea of Fertility doesn't stand out as well here as it does back on Earth. There's not as much contrast between that and the surrounding craters."

He paused for a moment, taking in more of the expanse beneath him.

"The craters are all rounded off. There's quite a few of them, some of them are newer. Many of them look like, especially the round ones, look like [they were] hit by meteorites or projectiles of some sort. Langrenus is quite a huge crater; it's got a central cone to it. The walls of the crater are terraced, about six or seven different terraces on the way down."

A few minutes later, the spacecraft passed over one of the sites NASA had identified as a potential landing area for future missions.

"It's very easy to spot," Lovell said. "You can see the entire rims of the craters from here with, of course, the white crescent on the far side where the Sun is shining on it."

A few seconds later, Borman jumped in. He still couldn't believe the accuracy with which planners had calculated the flight.

"Houston, for your information, we lost radio contact at the exact second you predicted. Are you sure you didn't turn off the transmitters at that time?"

"Honest injun, we didn't," Carr replied.

"While these other guys are all looking at the Moon, I want to make sure we got a good SPS," Borman said. Without the properly functioning SPS engine, Borman knew, Apollo 8 could never leave lunar orbit.

"And we want a Go for every rev, please," Borman added. "Otherwise, we'll burn in TEI-1 at your direction."

It was the request of a conservative commander. Before every new orbit, Borman wanted Houston to confirm that everything— the spacecraft, its systems, its computers—was working well. If not, he was prepared to fire his engine, leave lunar orbit, and head home—Trans Earth Injection, or TEI—at the first opportunity. In Borman's voice, even from a distance of a quarter million miles, Kraft could hear he had the right commander on board.

Apollo 8 continued flying, more and more nose-first, over the near side of the Moon. Inside, Anders kept his still and movie cameras firing, trying to record as much of the lunar surface as possible, all according to the photographic plan provided by Houston. Aiming and focusing weren't easy. The center window had been fogged by sealant fumes. Framing panoramas from the small rendezvous window was like trying to look out over the Grand Canyon through a welder's helmet. And when Anders did lock on to something good, he might have to interrupt the moment to change lenses or swap out film magazines. Still, as the Moon moved under the spacecraft, Anders began to capture spectacular shots, hundreds of close-up answers to questions that had endured for millennia.

Lighting conditions stayed good for another few minutes before Apollo 8 flew into darkness. (Generally, the crew would have about an hour of good lighting for photography during each two-hour orbit.) Forty minutes later, the ship slipped around to the lunar far side, where it again lost contact with Earth. Apollo 8 had made its first full revolution; when they next came around, they would be making their first TV broadcast from the Moon. It would be an early morning telecast in the United States, but millions would be watching and listening, there and all over the world. At home, the astronauts' wives gathered their children in front of their television sets. None of the women had been able to sleep.

At around 7:30 A.M. on December 24, test patterns flickered on TV screens and a grayish blob wobbled into the picture. When the camera steadied, the blob settled into a perfect sphere, with faint, almost invisible circles etched onto its surface, or maybe they were just lines, or the viewer's imagination. But when Anders pointed the camera out a window with better visibility, even the youngest viewers knew what had come into their homes.

This was the Moon.

"Say, Bill," Lovell said, playing emcee for the broadcast, "how would you describe the color of the Moon from here?"

Much of the world might have expected a poetic description. But as Anders looked down, the lunar surface reminded him of the seawall at La Jolla Shores in San Diego, where he and Valerie used to roast marshmallows and play volleyball when they were younger. So that's how he described it.

"The color of the Moon looks a very whitish-gray, like dirty beach sand with lots of footprints in it."

Flying past various landmarks, Anders worked the camera for a clearer view. After a time, the picture became sharp. One after another, Anders not only described the craters he was seeing, but referred to them by names that he himself had bestowed. He reported that Apollo 8 had passed over Mueller, Bassett, See (Bassett and See were two astronauts who'd died in an airplane crash in 1966), Borman, Lovell, Anders, Collins, and others.

In Moscow, when the news emerged that Apollo 8 had made its lunar orbit, the reality of the moment struck cosmonaut Alexei Leonov, who was training to command a Soviet circumlunar mission. Watching the astronauts, he felt his life's work crumble, his dreams evaporate. He worried that with news of Apollo 8, the Soviet Union might scrap its entire manned circumlunar program. Yet he could not but help respect the Americans, not just for what they'd done, but how they'd done it. To him, the aggressive, last-minute upgrade of Apollo 8 was nothing short of inspired.

Twelve minutes into the broadcast, Apollo 8 signed off and television screens went dark. Even with the cameras off, the astronauts couldn't stop describing the Moon.

"The view at this altitude, Houston, is tremendous," Lovell told CapCom Jerry Carr. "There is no trouble picking out features that we learned on the map."

Moments later, Lovell arrived at a place he'd long been waiting to reach.

"I can see the old second initial point here very well—Mount Marilyn."

"Roger," Carr confirmed.

On Earth, Lovell had promised his wife he would name a mountain for her. Now, from the Moon, he'd delivered.

An hour after passing Mount Marilyn, Apollo 8 disappeared again behind the western limb of the Moon to complete its second revolution. The astronauts now had to prepare to fire their SPS engine again, this time for just a few seconds, to circularize their orbit. Until now, they'd been flying an ellipse, one ranging from about 69 miles to about 195 miles above the lunar surface. A successful burn would put them at a constant altitude of 69 miles.

As before, a display flashed 99, the crew pushed a button to proceed, and the engine lit. Eleven seconds later, it stopped. By the calculations of the onboard computer, Apollo 8 was now in a circular orbit about 69 miles above the Moon. And it would stay there, long after the astronauts were dead, unless the SPS engine fired again, an event scheduled in sixteen hours, and about which Borman was growing increasingly apprehensive.

As Apollo 8 orbited the Moon, Borman was in charge of piloting the craft. In an airplane on Earth, that meant steering, turning, changing altitudes, and a host of other operations. At the Moon, where Apollo 8's path was locked in by orbital mechanics, it meant keeping the spacecraft oriented so that his crewmates could carry out their duties. Until now, the ship had been flying backward, necessary in order for the SPS engine to slow the craft enough to put it into lunar orbit and then to circularize that orbit. But now, following the flight plan, Borman began to pitch the spacecraft downward until Apollo 8 pointed nose down and vertical to the Moon. With the new view, Anders could begin shooting a series of vertical stereo photographs— two photos of the same object from slightly different positions— that would aid NASA in constructing detailed topographic maps of the lunar surface, including the approach path for future landing missions. He continued to concentrate on photography whenever there was light to shoot, as well as on monitoring the spacecraft and its systems.

Lovell continued to study the lunar terrain and take sightings and photos of lunar landmarks. After centering an important place or feature (many of which NASA had preselected) in the optics of his sextant, he would push a button on a control panel in front of him that recorded the spacecraft's location and the exact angle to the landmark. Collecting the precise coordinates of these places would help NASA build more detailed maps of the Moon, refine their knowledge of its shape, and chart variations in its gravity field that might draw future missions off course.

By now, the flight was just over three days old, and none of the crew had found much rest, another problem in Borman's file cabinet of concerns. Apollo 8 was scheduled for just ten orbits, and the third one had already started. And that was the rub. How could a man come to the Moon for just twenty hours and spend any of it snoring in his hammock? And yet Borman believed that if the crew didn't rest, mistakes would be made, some of them potentially catastrophic. But when he looked around the cabin, all he could see was Lovell and Anders busily at work.

Anders was immersed in his cameras when Apollo 8 came around for its third pass across the lunar near side. It was difficult for the astronauts to estimate the dimensions of the craters and mountains they were seeing, or even gauge that the spacecraft was at an altitude of 69 miles. When flying in an airplane, an observer sees familiar reference points—a city block, a river, an automobile—that help determine altitude, distance, even speed. Flying over the Moon, the astronauts saw only craters and more craters, and mountains in between. Without their knowing the size of those craters and mountains, any sense of distance or altitude was short-circuited. By reasoning, the men knew they weren't, say, one mile above the Moon, because the surface wasn't whizzing by beneath them. But much beyond that, it was hard for them to be certain of anything by means of the naked eye.

Two hours, and a full revolution, later, Borman still had the spacecraft pointed nose down. The position gave the astronauts their clearest view yet of the lunar surface. To each of them, the Moon appeared a place of sameness and loneliness, an expanse of blacks and whites and grays.

With four minutes remaining until Apollo 8 emerged from the eastern limb and reestablished contact with Earth, Borman fired his thrusters and put the ship into a 180-degree roll to the right, just as the flight plan dictated, so that Lovell could take sightings of lunar landmarks. The spacecraft was still pointed nose down, but for the first time since arriving at the Moon, the windows faced forward, in the direction of travel.

In the distance, the astronauts could see the arc of the lunar horizon, and beyond it, the pitch-black infinity of space. As Apollo 8 continued to roll, Anders saw something appear in his window, just over the Moon's western horizon.

"Oh, my God!" he called out. "Look at that picture over there! Here's the Earth coming up. Wow, is that pretty!"

A shining sphere of royal blues, swirling whites, and dabs of sunbaked browns rose over the rough, all-gray Moon. And now Borman and Lovell saw it, too.

Anders reached for his camera.

"Hey, don't take that, it's not scheduled," Borman joked.

But no one could take his eyes off the scene.

"Hand me that roll of color, quick, will you?" Anders said.

"Oh, man, that's great!" Lovell said.

"Hurry, quick!" Anders said, as Earth continued to rise above the horizon. In a few moments, he knew, it would be gone.

And then Earth disappeared.

"Well, I think we missed it," Anders said, his voice soft, his disappointment palpable.

"Hey!" Lovell cried several seconds later, looking through the hatch window. "I got it right here!"

The spacecraft was still rolling. The scene had shifted windows. Earth was still rising, and it looked brighter than ever.

"Let me get it out this window," Anders said, looking through his rendezvous window. "It's a lot clearer."

By now, Anders had swapped out his black-and-white film for color. Armed with his Hasselblad 500 EL still camera and Zeiss Sonnar 250 mm telephoto lens, he fired off the first color shot of Earth, now clearly above the lunar horizon.

"You got it?" Lovell asked.

Anders confirmed it.

"Well, take several of them," Borman said.

Lovell could hardly contain himself—Earth was retreating from him as the spacecraft continued to move.

"Take several of them! Here, give it to me."

"Wait a minute," Anders said. "Let's get the right setting here now. Calm down, Lovell."

"Well, I got it right . . . Aw, that's a beautiful shot," Lovell sighed.

Anders adjusted the exposure on his camera, then took another color picture.

"You sure we got it now?" Lovell asked. He still could not quite process what he'd seen.

"Yes," Anders replied, smiling. "We'll . . . it'll come up again, I think."

A moment later, the spacecraft rolled so far that Earth finally vanished from its windows.

The men were due to reestablish contact with Houston in just one minute. For now, no one said a word.

Earthrise was the most beautiful sight Borman had ever seen, the only color visible in all the cosmos. The planet just hung there, a jewel on black velvet, and it struck him that everything he loved—Susan, the boys, his parents, his friends, his country—was on that tiny sphere, a brilliant blue and white interruption in a never-ending darkness, the only place he or anyone else had to call home.

Lovell was overwhelmed by the smallness of Earth, home to three and a half billion people who, from this vantage point, all wanted the same things—a family to love, food to eat, a roof over their heads, children to kiss. From this distance, he could scarcely comprehend the fragility of Earth's atmosphere, a layer no thicker than the skin on an apple, the only thing that protected those lives, and life itself.

To Anders, Earth appeared as a Christmas tree ornament, hung radiant blue and swirling white in an endless black night. From here, it was no longer possible to pick out countries or even continents; all a person could see was Earth, and it occurred to Anders, in this last week of 1968, this terrible year for America and the world, that once you couldn't see boundaries, you started to see something different. You saw how small the planet is, how close all of us are to one another, how the only thing any of us really has, in an otherwise empty universe, is each other. As Apollo 8 came around the limb of the Moon and readied to reconnect with home, it seemed to Anders so strange—the astronauts had come all this way to discover the Moon, and yet here they had discovered the Earth.

Chapter Twenty

●

THE HEAVEN AND THE EARTH

AS APOLLO 8 MOVED THROUGH ITS FOURTH PASS OVER THE near side, NASA's public affairs officer provided sundry statistics for the media, as he did periodically. The spacecraft was traveling at 3,560 miles per hour; Anders's recent average heart rate had been 68 beats per minute, with a high of 69 and a low of 67; cabin temperature was 79 degrees Fahrenheit, two degrees warmer than an hour ago; cabin pressure was 4.9 psi. All of this looked normal to Mission Control.

One statistic that did concern Houston was sleep. When Cap-Com Mike Collins radioed for a status report, Borman acknowledged that the crew had managed only a couple of hours' rest over the last sixteen hours or so.

In fact, no one had gotten much sleep during the entire flight, which had now lasted three days, four hours, and change. As long as no one was resting, Collins radioed the latest news from Earth:

"We got the Interstellar Times here, the December twenty-fourth edition. Your TV program was a big success. It was viewed this morning by most of the nations of your neighboring planet, the Earth. It was carried live all over Europe, including even Moscow and East Berlin. Also in Japan and all of North and Central America, and parts of South America. We don't know yet how extensive the coverage was in Africa. Are you copying me all right? Over."

"You are loud and clear," Borman answered.

"Good," Collins continued. "San Diego welcomed home today the *Pueblo* crew in a big ceremony. They had a pretty rough time of it in the Korean prison. Christmas cease-fire is in effect in Vietnam, with only sporadic outbreaks of fighting. And if you haven't done your Christmas shopping by now, you better forget it."

As Apollo 8 streaked over the lunar surface, newspaper reporters on Earth moved just as fast to feed the public's insatiable appetite for astronaut stories. One article in *The New York Times* focused solely on the fact that each crewman was an only child. Another noted that Pan American World Airways had been inundated with requests from customers who wanted to reserve a seat on the first commercial flight to the Moon. (So far, the airline had about a hundred names on the waiting list.)

Apollo 8 passed behind the Moon in total darkness, just as it had when it arrived. When it came back around to the near side for its fifth revolution, Lovell suggested that Borman get some sleep. The commander was due for three hours of rest, and he tried to take it (though he wouldn't risk another sleeping pill). Lovell and Anders continued their work but grew frustrated with the limited visibility on account of hazing caused by the Sun. It didn't stop them, however, from continuing to watch the Moon.

"It doesn't seem like we've hardly been here that long, does it?" Anders said.

Lovell recalled his childhood, when he'd dreamed of an opportunity like this.

"It seems like I've been here forever," he said.

"You know," Anders remarked, "it really isn't all that . . . anywhere near as interesting as I thought it was going to be. It's all beat up."

"The things that I saw that were interesting were the new craters," Lovell said. He liked the idea that the Moon remained alive in the heavens, that it was still changing, still becoming.

A few minutes later, the spacecraft slid again behind the lunar far side. Apollo 8 had now been at the Moon for about ten hours and was halfway through its ten orbits. Just ten more hours remained until Trans Earth Injection, or TEI, the maneuver designed to get Apollo 8 out of lunar orbit and on its way back home. Nothing worried Kraft, and many others at NASA, more than TEI. So much could go wrong, and with such dire consequences. The men back in Houston tried to remain optimistic. Around the time Apollo 8 disappeared behind the Moon (about three o'clock in Houston), a message lit up on one of Mission Control's large data panels. In red, white, and blue letters, it read MERRY CHRISTMAS APOLLO 8.

By the time Apollo 8 launched, NASA was considering just two possible sites for a future landing mission. Both were located in the Sea of Tranquillity, to the right side of the full Moon as seen from America and other places in Earth's northern hemisphere. NASA wanted to land during the lunar morning, when temperatures were moderate and low Sun angle would create long shadows that would help a commander discern a smooth spot on which to set down. But those conditions shifted every day on the Moon. By choosing two sites, twelve degrees apart, NASA ensured that if it had to delay launch by a day, the lunar module would still have an optimal landing site when it arrived.

Both sites also satisfied other important NASA criteria for the first lunar landing. They were accessible to a spacecraft flying a free-return trajectory—a NASA safety requirement—and they existed in areas with ample level terrain, which meant a lunar module wouldn't have to expend an undue amount of propellant hovering and maneuvering to avoid boulders and slopes before setting down.

Among Apollo 8's tasks were to confirm that its own trajectory could be used by future spacecraft to reach these landing sites, and to get a close-up view of the areas under the same lighting conditions as the future landing mission would encounter. As Apollo 8 coasted over the first of these sites during its sixth pass over the near side, Lovell described it for Houston. Even the shadows, a critical element to judging shape, depth, and distances, looked excellent to him.

"I have a beautiful view of it. The first [landing site] is just barely beneath the vertical now, and the second one is coming up—it's just a grand view."

As the spacecraft moved over the second landing site, Lovell yearned to set down there; it seemed as close as the aircraft carriers he'd landed on so many times. He told himself he would come back here, not just to observe the Moon, but to walk on it.

Just before Apollo 8 slipped behind the Moon for its seventh pass over the far side, Lovell began singing to himself, as was his habit, then turned to his crewmates.

"Did you guys ever think that one Christmas you'd be orbiting the Moon?"

"Just hope we are not doing it on New Year's," Anders replied, his wit growing drier with each orbit. There was a dark truth behind Anders's humor. If Apollo 8 was still here in a week, it meant the crew was never coming home.

Susan Borman knew it, too. She cleared her kitchen table, sat, and started to compose Frank's eulogy. She needed to be ready—not like her friend Pat White, who'd been taken by surprise by the death of her husband in the Apollo 1 fire, and by the swiftness with which government officials moved in to orchestrate funeral arrangements. This time, Susan would be in charge. She would do it the way she and Frank wanted it, and the way that was right for their sons. It seemed to her a better fate for a man like Frank to die in space than to burn up on the launchpad while training, and a better fate for her, knowing Frank was in a place he'd be forever, a beautiful Moon she could see in the night, a place where she could always find him.

Just eight and a half hours remained before Trans Earth Injection. On board Apollo 8, Anders secretly hoped something would go wrong—nothing catastrophic, of course, just enough that he could show Houston, and his crewmates, how beautifully he'd mastered the spacecraft and its systems. But the ship was proving to be a jewel.

As the spacecraft readied to reconnect with Houston and begin its seventh pass across the lunar near side, Borman called out to his crewmates.

"Oh, brother! Look at that!"

"What was it?" Lovell asked.

"Guess," Borman said.

Lovell did some quick computations. The ship was above the far side, at around 120°E longitude, and at the most southerly part of its orbit. For Borman to react like that, he must have seen Tsiolkovsky, one of the far side's most impressive craters, 115 miles wide, with a peak rising 2 miles out of its sunken center, and 80-foot boulders strewn about. So that's what Lovell guessed.

"No," Borman said. "It's the Earth coming up."

Through his window, Borman had caught another Earthrise, this one as stunning as the first, not just for its beauty, but for how it came to him—unexpected, ascendant, a call from home.

In Houston, Marilyn Lovell felt the need to go to church. Late night Christmas Eve services weren't scheduled to start for several hours, but Father Raish told her to drop by anyway. When she arrived late that afternoon, the church was decorated with flowers and Christmas trimmings and burning candles. Marilyn was the only parishioner there. While the church organist played, Marilyn took a private communion, then joined Father Raish in prayer—for Jim, for his crewmates, for the mission. In just a few hours, they knew, Apollo 8 would face perhaps its most dangerous and critical test. And it would all happen just a few minutes after midnight on Christmas morning.

Only seven hours remained until Trans Earth Injection. But before the crew could get ready for that, they had to prepare for their second television broadcast from the Moon. It would occur in less than four hours, at around 8:30 P.M. Houston time, on Christmas Eve, before children's bedtimes in America. By NASA's estimates, more people around the world would be watching and listening than had ever tuned in to a human voice at once.

These last few hours demanded the best of the crew. The Apollo spacecraft was incredibly complex to operate, and the SPS engine was no exception. For Trans Earth Injection, there were five pages of switch settings, equipment checks, and adjustments, each of which had to be verified by a second crewman in the knowledge that one mistake could prove fatal.

But as Borman looked around the cabin, he doubted that he'd be getting the best from the crew. None of them had slept for the past eighteen hours. Each was starting to get sloppy, miss things, make mistakes. And mistakes, Borman knew, had a way of spiraling into catastrophe.

Borman told Houston he was scrubbing most of the flight plan for the next orbit.

"We are a little bit tired," he told Collins. "I want to use that last bit to really make sure we're right for TEI."

On board, he made it clear to the others what he wanted.

"You're too tired, you need some sleep, and I want everybody sharp for TEI," he said.

Borman seemed exasperated with everybody, even the flight planners, who had loaded up the final orbits with tasks for the astronauts.

"Unbelievable!" Borman said. "The detail these guys study up. A very good try but just completely unrealistic, stuff like that. I should have—"

"I'm willing to try it," Anders said, offering to perform some of the duties that Borman wanted scrubbed. But Borman wouldn't hear of it.

"I want you to get your ass in bed! Right now! No, get to bed! Go to bed! Hurry up! I'm not kidding you, get to bed!"

It was a conversation many would be having in their own households this night, Christmas Eve.

Lovell and Anders kept talking about cameras and lenses instead of immediately obeying their commander. A few minutes later, Apollo 8 disappeared behind the lunar far side and lost contact with Earth. Lovell had finally gone to his hammock, but Anders was still on duty.

"Goddammit, go to bed!" Borman told him.

Anders offered a counter, and Borman exploded.

"Get going! I think this is a closed issue. Get to bed! . . . No, you get to bed, get your ass to bed. You quit wasting one . . . I don't want to talk about it . . . Shut up, go to sleep, both you guys! . . . You should see your eyes—get to bed! . . . Don't worry about the [camera] exposure business, goddammit, Anders, get to bed! Right now! Come on! . . . You've only got a couple hours, Bill, before we're going to have to be fresh again."

Borman clenched his teeth. In six hours, everything had to work perfectly, but all of them were getting groggy and sloppy. In his hammock, Lovell began to fear that the strain of command had finally overcome Borman. But Borman never wavered in his insistence that his crew go to sleep. A few minutes later,

Lovell and Anders had gone quiet and were finally resting. It was a military chain of command, even at the Moon.

When Apollo 8 regained contact with Houston, there were just two hours—one revolution—remaining before the big television broadcast. CapCom Ken Mattingly asked for an update on the crew's sleep. Borman reported that Lovell and Anders were currently resting, and that he'd had "about three or four hours earlier today." In fact, Borman had managed only eighty minutes' rest since arriving at the Moon.

During this coast over the near side, Borman radioed for a weather report on Earth. Mattingly reported that all looked well, including in the Pacific, where Apollo 8 would splash down.

"They told us that there is a beautiful Moon out there," Mattingly said.

"Now, I was just saying that there's a beautiful Earth out there," Borman replied.

A few minutes later, Borman asked, "Hey, Ken, how'd you pull a duty on Christmas Eve? You know, it happens to bachelors every time, doesn't it?"

"I wouldn't be anywhere else tonight," Mattingly replied.

The astronauts' wives had settled in for the historic television broadcast, scheduled for eight thirty that evening. The astronauts' children seemed calmer than their mothers and the family friends, and more focused on Christmas presents than on Christmas orbits.

Two of the wives chose to be home with family and friends. Marilyn, who'd never before been apart from her husband for Christmas, was asking the crowd inside her family room, "Why is everyone here? You should be home with your families!" but even though it was Christmas Eve, not a single person would leave. At her home, Valerie, who'd been apart from Bill only once during Christmas, in 1957, when he was stationed in Iceland and she was home with two babies, was bringing doughnuts and coffee to the journalists and state troopers massed on the front lawn, and getting her kids ready to listen to their dad.

Only, Susan Borman was not at home. To find respite from reporters and commotion, she'd taken her sons to a friend's house in Houston to eat dinner and watch the telecast in peace. There, she thought back to 1951, to the only time she and Frank had ever been separated at Christmas.

Frank had been ordered to the Philippines for fighter pilot training. On Christmas Eve, his transport ship stopped in Hawaii. He'd never missed Susan more than on that night, nor she him. Susan presumed she wouldn't hear from Frank for days, but he resolved to find a phone. Only the Royal Hawaiian Hotel had one in the vicinity, so Frank went into the phone booth in the lobby, dialed the operator, and asked to place a call to Arizona. "I'm sorry, all the phone lines have been booked for months," the operator told him. Years later, Frank would remember this as the lowest moment of his life to that point. A well-dressed gentleman must have noticed the expression on his face. "What's wrong, Lieutenant?" he asked. Borman explained. The man introduced himself as the hotel manager, then gave Frank a key and told him to use the phone in one of the rooms. "Talk as long as you want," he said. Frank offered money but the man refused. Soon, Frank and Susan were talking and saying "I love you" and wishing each other a merry Christmas. Their call lasted for more than an hour. Now, seventeen years later to the day, Susan remembered that call, and how close she had felt to Frank despite their distance, and from outer space, Frank remembered it, too.

One of Apollo 8's objectives was to investigate the effects of mass concentrations, or "mascons," on a spacecraft's orbit. These areas of increased density in the Moon's crust, primarily caused by massive asteroid impacts, subtly altered a ship's trajectory (by changing the gravity field) and, if not compensated for, would eventually cause it to crash into the lunar surface. As it looked now to Mission Control, the mascons were detectable, but their effect on Apollo 8 was slight. Future lunar modules, however, would be flying much lower, and be more subject to their influence.

As the spacecraft traveled yet again behind the Moon, one hour remained until the broadcast. Even now, the astronauts weren't sure how they wanted to run it.

"I don't think we ought to screw around with this," Borman told his crewmates. "We've got to do it up right because there will be more people listening to this than ever listened to any other single person in history."

They had long known that they would need words worthy of the moment. The astronauts had tossed around ideas in the weeks leading up to the flight, but none had seemed appropriate. They considered telling a Christmas story, but the flight was important not just to Christians but to all faiths, and to humanity. They thought about invoking Santa Claus, but that didn't seem serious enough for such a historic occasion. Changing the words to "Jingle Bells" was silly. But they'd had no better ideas. And time was running out.

In early December, with just two weeks remaining until launch, Borman had asked a friend, a sensitive and intellectual man named Si Bourgin, for help. Bourgin put his mind to the problem but wasn't happy with his ideas. In turn, he approached Joe Laitin, a former war correspondent, and gave him twenty-four hours to find the right words. Laitin worked deep into the night, also to no avail. At 3:30 A.M., Laitin's wife, Christine, made a suggestion. It was her idea that was forwarded to Borman, who showed it to his crewmates. All three astronauts agreed that they'd found the right message. From that point forward, the men spoke about it to no one, not even their wives.

Inside Mission Control, every square foot was packed with NASA personnel. At their homes in Houston, the astronauts' wives gathered around television sets with children, friends, and family, gifts for their husbands wrapped and placed under twinkling Christmas trees, awaiting their return. In sixty-four countries, a billion people—more than one-quarter of the world's population—joined them, pushing close to their own televisions and radios, waiting to hear what the first men at the Moon would say on Christmas Eve.

At 8:30 P.M. Houston time, CBS cut away from *The Doris Day Show* to Walter Cronkite, and other American television networks also interrupted their normal programming. Four minutes later, dark horizontal lines wobbled on viewers' screens. A small, bright orb shone in the upper left part of the picture—likely Earth, but no one could tell for sure.

"This is Apollo 8, coming to you live from the Moon," Borman said.

He explained to viewers how the crew had spent Christmas Eve—doing experiments, taking photographs, firing their thrusters—and promised to show everyone a lunar sunset. But first, he wanted to talk about the place he and his crewmates had been circling for the past sixteen hours.

"The Moon is a different thing to each one of us. I think that each one of—each one carries his own impression of what he's seen today. I know my own impression is that it's a vast, lonely, forbidding type existence, or expanse, of nothing. It looks rather like clouds and clouds of pumice stone, and it certainly would not appear to be a very inviting place to live or work. Jim, what have you thought most about?"

"Well, Frank, my thoughts were very similar," Lovell said. "The vast loneliness up here of the Moon is awe-inspiring, and it makes you realize just what you have back there on Earth. The Earth from here is a grand oasis in the big vastness of space."

Anders chimed in.

"I think the thing that impressed me the most was the lunar sunrises and sunsets. These, in particular, bring out the stark nature of the terrain, and the long shadows really bring out the relief that is here and hard to see in this very bright surface that we're going over right now."

Suddenly NASA lost the picture from Apollo 8, and so did the rest of the world. But the audio remained clear, and Anders continued to describe some of the landmarks he was seeing as Mission Control struggled to regain the visual. Soon the picture returned, this time a view out a different window, one that showed the clear arc of the grayish-white Moon against the pitch-black lunar horizon. Anders described the various craters he could see as the spacecraft glided overhead.

"Actually, I think the best way to describe this area is a vastness of black and white, absolutely no color," Lovell said.

"The sky up here is also a rather forbidding, foreboding expanse of blackness, with no stars visible when we're flying over the Moon in daylight," Anders added.

For the next several minutes, the astronauts continued to describe what they were seeing—mountains, craters, landmarks, the brilliance of the Sun's reflection. At one point, Anders became so enthused about describing the evolution of craters that Borman had to remind him, off air, "Hey, Bill, you're not talking to geologists." Anders changed windows for a better view, only to have the audio nearly overcome by static as the spacecraft flew over the Sea of Crises. Soon, however, things cleared up near the Sea of Fertility.

"How's your picture quality, Houston?" Anders asked.

"This is phenomenal!" CapCom Mattingly replied.

"We're now going over . . . approaching one of our future landing sites," Anders said, "selected in this smooth region to . . . called the Sea of Tranquillity . . . smooth in order to make it easy for the initial landing attempts in order to preclude having to dodge mountains. Now you can see the long shadows of the lunar sunrise."

The scheduled television time was winding down, and there was one important thing left to do. As the spacecraft moved across the Sea of Tranquillity, Borman motioned to Anders.

"We are now approaching lunar sunrise," Anders said, "and for all the people back on Earth, the crew of Apollo 8 has a message that we would like to send to you."

No one at Mission Control, or anyone else, had any idea what these men were about to say.

The astronauts' wives and children and friends leaned forward.

While the Moon continued to move across television screens, Anders began.

"In the beginning, God created the heaven and the earth. And the earth was without form, and void, and darkness was upon the face of the deep. And the spirit of God moved upon the face of the waters. And God said, 'Let there be light.' And there was light. And God saw the light, that it was good, and God divided the light from the darkness."

Anders was reading the first words from Genesis, the first book of the Bible.

Lovell continued the passage.

"And God called the light Day, and the darkness he called Night. And the evening and the morning were the first day. And God said, 'Let there be a firmament in the midst of the waters, and let it divide the waters from the waters.' And God made the firmament and divided the waters which were under the firmament from the waters which were above the firmament. And it was so. And God called the firmament Heaven. And the evening and the morning were the second day."

Borman continued.

"And God said, 'Let the waters under the Heaven be gathered together unto one place. And let the dry land appear.' And it was so. And God called the dry land Earth. And the gathering together of the waters He called seas. And God saw that it was good."

Borman paused.

"And from the crew of Apollo 8, we close with good night,

good luck, a Merry Christmas, and God bless all of you—all of you on the good Earth."

A moment later, television screens around the world went dark.

Inside Mission Control, no one moved. Then, one after another, these scientists and engineers in Houston began to cry.

The agency had allowed Borman to choose what to say to the world on Christmas Eve—no oversight, no committees, not even a quick glance on the day before the flight departed. It had come as a complete surprise to them.

In his studio at CBS, Walter Cronkite fought back tears as he came back on the air.

At a house party in Connecticut, novelist William Styron told himself to remember the scene. He had had to persuade his host, the composer Leonard Bernstein, to watch the broadcast. Bernstein considered the space program an overhyped waste of vast American treasure, but he'd bent to the wishes of his guest. As the astronauts read from Genesis, the raucous party went still. Styron would never forget the emotion on Bernstein's face during Borman's parting words, a look he would describe years later as "depthless and inexpressible."

Watching in Houston, Susan Borman wept. Marilyn Lovell gathered up her kids and they walked, not drove, past the holiday lights in Timber Cove, slow enough to remember them all. Valerie Anders told her children, "That was for the whole world."

Across much of the globe, people streamed outside and looked up, trying to pick out the three men who'd just spoken to them, knowing it was impossible, but trying all the same.

Chapter Twenty-One

●

AIMING FOR HOME

ON BOARD APOLLO 8, BORMAN WASN'T CERTAIN THAT THE crew's message had even been heard.

"Did you read everything that we had to say there?" he radioed to Houston.

"Loud and clear," Mattingly confirmed. "Thank you for a real good show."

That settled, Borman got down to the matter that had concerned him for months: Trans Earth Injection. Perhaps more than any other part of the mission, Trans Earth Injection, or TEI, had haunted NASA managers, planners, and controllers. Without it, Apollo 8 could not return home.

Since entering lunar orbit, the spacecraft had been traveling steadily at about 3,600 miles per hour. And unless it could gain enough speed to overcome the lunar gravity holding it in orbit, it would never leave the Moon. To do that, Apollo 8 would need to increase its speed to about 6,000 miles per hour. Onboard thrusters weren't nearly powerful enough to provide that kind of boost. Only the SPS engine—the same one the crew had used to enter lunar orbit—had the muscle it would take.

As before, the engine needed to burn for just the right amount of time, with just the right amount of thrust, and in the right direction, to send the spacecraft and its crew home safely. If it burned too long or too strong (or both), Apollo 8 could be hurled off into space without enough propellant to correct the bad trajectory and set course back to Earth, and would be doomed to pursue its own orbit around the Sun. If it burned too short or too weak (or both), the spacecraft could coast off into space without sufficient momentum either to return to the Moon or to fall back to Earth, or it might crash into the lunar surface, adding another crater to the Moon.

But perhaps the worst result would come if the engine failed to light at all. In that case, Apollo 8 would remain a possession of the Moon for eternity.

NASA had confidence in the SPS engine. While it had not performed well in its brief test firing at eleven hours into the flight, it had functioned twice without issue at the Moon, when it was used to enter, and then to circularize, lunar orbit. Still, that was no guarantee. With the TEI maneuver just three hours away, Kraft's dread began mounting. Rockets, he knew, were complex, temperamental, violent machines. They failed, blew up, or shut down, often without warning. In the harsh environment of space, a rocket's highly pressurized mélange of moving mechanical parts, which generated intense friction and heat and depended on proper lubrication and coolants and a universe of electrical connections, could go wrong in any number of ways.

While the astronauts worked to pack away loose items and prepare for TEI, Mission Control examined data from the spacecraft and its systems to determine whether Apollo 8 was ready. They broadcast their decision seventy-one minutes later, after the ship had emerged from the Moon's eastern limb.

"Okay, Apollo 8," Mattingly radioed to the crew. "We have reviewed all your systems. You have a Go for TEI."

The maneuver was now just one hour and fifteen minutes away. The crew began its final preparatory procedures, running down checklists while straining to keep the bright Sun from their eyes. Much of the exchange between the men was technical and rote, indecipherable to a lay listener, but comforting in the way its call-and-response rhythms sounded like a preacher and his congregation:

Anders: Okay, let's go to P40: P30, complete; CMC, On.

Borman: CMC is On.

Anders: ISS, On; spacecraft SCS, operating.

Borman: Right.

Anders: Test the Caution/Warning lamp; EMS mode, Standby.

Borman: Yes.

Anders: Function, delta-v set.

Borman: Right.

Anders: And have you set 1586.8?

Borman: Right.

Anders: Okay, EMS mode, Standby; delta-v set; set delta-vc.

Borman: 3501.8.

Lovell: I'll check: 3501.8.

Borman: Okay.

Anders: EMS Function, delta -V; Nonessential Bus, Main B; cycling cryo fans—good a time to do it as any.

Anders: BMAG Mode, three, Rate two.

Borman: Rate two.

Lovell: Delta-*v*cg CSM.

It was now past eleven o'clock in Houston. Less than an hour remained until TEI. If all went according to plan, the SPS engine would fire at 12:11 A.M. Houston time, in the earliest minutes of Christmas morning.

At the Borman home, Susan and her sons would listen to the squawk box in front of a *Life* magazine reporter and photographer, and where NASA would know where to find them if something went wrong. After Valerie Anders put her children to bed, she drove over to the Borman house and joined Susan in the kitchen. Valerie and Susan liked each other, but they might have preferred to be alone at a time like this. NASA public affairs, however, had arranged this photo op, and the women went along with it. Marilyn had a touch of the flu and didn't attend.

As their lives in the public eye demanded of them, both women were dressed beautifully. Susan wore an ice-blue dress, a beige cardigan draped over her shoulders, a bangle bracelet on her wrist, and pearls around her neck. Valerie was in a robin's-egg-blue dress with scalloped rickrack at the neckline, a white cardigan embroidered with colorful flowers on her shoulders, and an elegant watch. Each had curled and styled her own hair (beauty shops were expensive on an astronaut's salary). A giant lunar map lay spread over Susan's table, a pack of cigarettes on top of that. At her own home, Marilyn stayed close to her squawk box and her friends, trying to stay as optimistic as Jim had seemed when he'd told her in August that everything would be okay.

As the wives settled in, Mission Control began to fill up with off-duty controllers, media, and others, until the place was again packed shoulder to shoulder, just as it had been when Apollo 8 first disappeared behind the Moon. Everyone there knew what to expect next.

At 11:42 P.M., Apollo 8 would begin its final scheduled pass behind the Moon, losing contact with Houston. Twenty-nine minutes later, while still out of communications, the SPS engine would fire, increasing the spacecraft's speed enough to leave its orbit and set course for Earth and its splashdown point in the Pacific Ocean. It would have been easier for the flight controllers had TEI occurred on the near side, where they could monitor the ship and talk to the crew, but orbital mechanics dictated that the break for home occur while Apollo 8 was on the far side. By design, the rocket would burn for about three minutes and eighteen seconds. If all went well, Houston could expect the spacecraft to emerge around the near side at about 12:19 A.M. If the rocket had malfunctioned, or had failed to fire, it would come out later than that, perhaps by as much as eight minutes. NASA possessed some of the world's most powerful computers, but it would be a simple clock that first told them whether their men were coming home.

Ten minutes now remained until Apollo 8 would disappear behind the lunar far side. Kraft and Low stood together in silence. If something happened to the astronauts—if the ship blew up or crashed into the Moon or flung itself off on an unrecoverable trajectory—NASA couldn't do anything about it, and they wouldn't even know about it until after it happened.

In 1961, Kraft had been the flight controller on Mercury-Redstone 2, the first planned launch of a hominid into space. The passenger was a chimpanzee named Ham, and Kraft had become attached to him. By the time the rocket launched, Kraft regarded Ham as crew, and he celebrated Ham's safe return.

In the seven years since, Kraft had felt a personal responsibility to every man who risked his life aboard a NASA spacecraft. But all of them had been just a few hours from home in case of emergency. Now Kraft was about to say goodbye to three extraordinary men he both liked and admired, powerless to help them when they were days away from Earth, when they might need help most.

The countdown to loss of signal went under a minute. Susan chewed on her pearls while Valerie took several deep breaths.

At 11:42 P.M. Houston time, Apollo 8 slipped behind the Moon, and radio contact with Earth went dead.

Each of the astronauts was ready to come home. For Borman, America's mission to beat the Soviets to the Moon wouldn't be complete until the crew had returned safely. For Lovell, making it back meant a chance to return to the Moon, not just to see it but to walk on it. Anders, who'd been so interested in lunar geology, had seen all there was to see of Earth's satellite and didn't think he'd overlooked anything during his ten times around. All three of them missed their families.

"It's been a pretty fantastic week, hasn't it?" Borman asked his crewmates.

"It's going to get better," Lovell said.

"I hope this baby holds out for another two and a half days," Anders said. "It sure has performed admirably, hasn't it?"

None of the men had dwelled on what awaited them if the SPS engine didn't perform. If test pilots and fighter pilots thought like that, they would never climb into a cockpit. But none of the men could say he hadn't thought about being marooned in lunar orbit, or how he'd spend his remaining time—perhaps four days—before dying. In fact, Borman had been asked about it before launch.

"I don't know how I'd want to spend my last days," he'd told reporters. "I think that's something you decide when it happens. If the engine doesn't work, we've had a bad day."

NASA had considered a plan for a lunar rescue mission should something catastrophic happen. It involved sending a single astronaut to the Moon in his own command and service module, atop his own Saturn V, which would stand ready to launch at Cape Kennedy. Once in lunar orbit, rendezvous and rescue would involve complex maneuvers that would also place the rescuing astronaut at risk. Such a contingency would add significantly to the agency's already massive budget. In the end, the idea was scrapped.

NASA hadn't bothered training the astronauts on how to handle being stranded at the Moon, or being flung off irretrievably toward the Sun, or any other hopeless scenario. It hadn't supplied them with a suicide pill or any other means of putting an end to their lives. But the crew knew how things would end for them. About a week after TEI failed, the canisters of lithium hydroxide used to purge exhaled carbon dioxide from the cabin would run out, causing the men to grow drowsy, fall asleep, and suffocate.

None of them intended to waste that week, though they did not discuss the matter aloud. Almost certainly, they would have continued to make observations of the Moon, providing as much detail as possible for Houston. They would also have continued to wear a biomedical harness, to give NASA and its doctors information about what happens as one meets his end in space. And they would have radioed home to say goodbye to their parents, wives, and children, and told them how much they loved them.

But they might not have waited to suffocate.

Satisfied that their work had been done, the crew likely would have decided together to shut down their communications, then vent the spacecraft by opening a pressure relief valve. Doing so would cause an immediate loss of oxygen in the cabin, a fast loss of consciousness, and a painless death.

For now, the astronauts could only hope that that wouldn't be necessary.

Twenty minutes remained until the SPS engine was scheduled to fire for TEI. Ordinary people might use this time to say something profound, or perhaps to bid their companions goodbye in case things went bad. But NASA had selected Borman, Lovell, and Anders for a reason: This rare breed of man could, at once, love his wife and children and life with all his heart, yet still climb atop an unproven rocket and fly to the Moon. So as the clock counted down to ignition, it wasn't mortality and love these men were discussing.

"Tell you one thing these flights are good for," Borman said. "An old fatty like me, I bet I've lost a lot of weight. I didn't eat much those first two days, and I didn't—didn't even get much to eat today."

"Pretty sunrise," Anders remarked.

Inside Mission Control, there was little anyone could do but wait. Soon, people began talking and milling about.

That made Kraft furious. He got on his intercom and told anyone who could hear him to shut up so that he could pray or do whatever the hell else he could dream up to make sure Apollo 8 came out on the other side of the Moon when it was supposed to.

A few minutes later, the clocks in Mission Control read midnight. It was now Christmas Day in Houston, much of America, and the world. No one had ever been farther from home on this important family day than Borman, Lovell, and Anders.

From their windows, which faced toward the lunar surface, it appeared to the astronauts as if they would be headed for trouble when the rocket lit.

"It looks to me like I'm going to burn right into the ground," Borman said.

But the men didn't have time to worry about that. They'd long since maneuvered the spacecraft to the attitude NASA had calculated. They had faith that the agency had gotten it right.

Just thirty seconds remained until TEI.

"Flight recorder going to record," Anders called.

He'd made this flight believing he had a one-third chance of dying. Trans Earth Injection had been a major part of that calculus.

"Stand by to start ullage," Lovell called.

Lovell believed that at certain points in life, a person just had to have faith.

"Two valves," Borman called.

Borman had come for America, because he believed it was the greatest country on Earth and he would have died in order to protect it.

In Mission Control, people could barely breathe.

It was this moment that had so shaken James Webb when he heard of Low's plan. It wasn't just that the mission allowed only four months' preparation rather than the usual year and a half, or that it required manning a rocket that had flown only twice (and experienced myriad problems the second time), or that the crew would have no backup if the SPS engine failed, or that so much would have to be done for the very first time. What had shaken Webb most deeply was the idea that if the crew of Apollo 8 were stranded in lunar orbit on December 25, no one would ever look at Christmas, or the Moon, the same way again.

Five seconds remained until Trans Earth Injection.

Inside the spacecraft, the number 99 flashed on a display, asking the crew for the go-ahead to light the SPS engine and begin the burn. If no one pressed the Proceed key, ignition would not occur.

Lovell looked to Borman, and Borman nodded.

Lovell reached forward.

He pressed Proceed.

And then there was only silence.

Chapter Twenty-Two

●

PLEASE BE INFORMED—THERE IS A SANTA CLAUS

SUSAN BORMAN, VALERIE ANDERS, AND MARILYN LOVELL HAD been told that if all went well, their husbands would regain contact with Houston at about 12:19 A.M., the moment their spacecraft came around the lunar far side. That was still five minutes away. Seconds had never passed so slowly.

It wasn't much easier for those at Mission Control, and especially for Chris Kraft. All he could do now was wait.

Now, just one minute remained until Apollo 8 was due to regain contact with Earth. Any longer than that, and it meant something had not gone according to plan.

In Australia, technicians at the Honeysuckle Creek tracking station made certain their antenna was pointed accurately. (NASA needed a station in Australia, and elsewhere around the world, to ensure that the spacecraft could be "seen" at all times no matter where the Earth was in its rotation.) At just the moment Mission Control expected to acquire a signal, Australia reported receiving one. A wave of excitement washed over the room, but Houston still had to confirm it.

CapCom Ken Mattingly called to the spacecraft.

"Apollo 8, Houston."

There was no answer.

Mattingly waited a full eighteen seconds, then called again.

"Apollo 8, Houston."

Still no answer.

Susan Borman and Valerie Anders were silent. There was no sound in the Borman home but for the squawk box, and their husbands' voices were not coming out of it. Everyone there—Susan, her boys, the visitors—were just waiting to hear an astronaut's voice, which was now overdue.

Twenty-eight seconds later, Mattingly tried again.

"Apollo 8, Houston."

There was only silence.

At her home, Marilyn Lovell told herself that Jim had said everything would be okay.

This time, Mattingly allowed nearly a minute to pass before making his next call.

"Apollo 8, Houston."

Again, nothing came back from the spacecraft. Mattingly tried a fifth time, forty-eight seconds later.

Still no answer.

Almost four minutes had passed since the ground station in Australia had picked up a signal from Apollo 8 from behind the Moon—an unthinkable delay.

Then, over the static and hiss of the radio connection, a voice came through to Mission Control.

"Houston, Apollo 8, over."

The voice was Lovell's.

"Hello, Apollo 8," Mattingly answered. "Loud and clear."

"Roger," Lovell said. "Please be informed—there is a Santa Claus."

"That's affirmative," Mattingly said. "You're the best ones to know."

At the Borman home, Susan and Valerie threw up their arms and shouted with happiness. A *Life* magazine photographer captured the moment—the purest expression of simultaneous joy and relief one might ever hope to see. At the Lovell home, Marilyn squealed with delight and laughed out loud. *What a perfect thing to say*, she thought. *What a perfect thing to say today.*

It was twenty-five minutes past midnight in Houston, Christmas morning.

Lovell's words had been inspired by an article he'd once read, originally penned in the *New York Sun* in 1897. It told of an eight-year-old girl named Virginia who had asked the newspaper, "Please tell me the truth, is there a Santa Claus?" A longtime editor there, Francis Pharcellus Church, answered the girl's question. Church had been a Civil War correspondent for *The New York Times;* standing on the front lines with the Union's Army of the Potomac, he'd seen the terrible things men could do to each other, how a country could lose its heart and its soul when it did battle with itself. But for several paragraphs, Church talked about the realness of love, generosity, devotion, and beauty, even if one couldn't always see them, and how that proved that Santa, too, was real. And so he replied to the girl: "Yes, Virginia, there is a Santa Claus."

Leaving the Moon, Lovell echoed Church's words, and sentiments, even if he hadn't intended to.

Planners at Mission Control were so thrilled—and relieved—to have Apollo 8 back in contact that no one asked the obvious question: Why had the crew taken so long to respond to Mattingly's calls? In fact, the explanation was simple: Anders had been so busy confirming shutdown of the SPS engine, and grabbing cameras to shoot photographs, that he had forgotten to activate the spacecraft's high-gain antenna, which broadcast their signal back to Earth. Once Anders pushed the button, Lovell was clear to reconnect with the world.

For all the confusion, the SPS engine had performed flaw-lessly. The silence the astronauts experienced after Lovell pressed the button to light the engine was due to the time the computer took to digest information and send instructions to open the valves. Though the silence lasted only a moment, it felt like years to Lovell.

Once the engine fired, the crew was treated to a singular view, one that even Stanley Kubrick couldn't have equaled with all the special effects in Hollywood. Outside their windows, as Apollo 8 picked up speed and moved out of its circular orbit, the men could see the Moon receding, growing smaller before their eyes.

For most of human existence, people's ideas about the Moon derived from their imaginations, religious beliefs, and unaided eyes. In 1609, the Italian scientist Galileo Galilei peered through his homemade telescope and observed distinct features on the lunar surface, an ancient place that, with the aid of this won-drous new instrument, had suddenly become new to man's eye.

Galileo wrote, "We certainly see the surface of the Moon to be not smooth, even, and perfectly spherical, as the great crowd of philosophers have believed about this and other heavenly bodies, but, on the contrary, to be uneven, rough, and crowded with depressions and bulges. And it is like the face of Earth itself, which is marked here and there with chains of mountains and depths of valleys." He made beautiful sketches of what he saw, using shadows near the Moon's terminator—the line that divides dark from light on the lunar surface—to pick out features unde-tectable to the naked eye, including craters. The dark parts of the Moon, Galileo theorized, were low-lying plains or dry seas, the bright parts mountains and highlands.

Now, more than three centuries years later, three men had become the first to see with their own eyes the detail that Galileo had observed through his telescope, and they knew the sketches he made had been perfect.

Kraft finally allowed himself to exhale. He could see Deke Slayton, the man at NASA in charge of astronaut training and crew selection, step forward to speak to the men aboard Apollo 8. Ordinarily, it was just the CapCom who did the talking, but this moment was extraordinary.

"Good morning, Apollo 8, Deke here. I just would like to wish you all a very merry Christmas on behalf of everyone in the Control Center, and I'm sure everyone around the world. None of us ever expect to have a better Christmas present than this one. Hope you get a good night's sleep from here on and enjoy your Christmas dinner tomorrow; and look forward to seeing you in Hawaii on the twenty-eighth."

"Okay, leader," Borman replied. "We'll see you there. That was a very, very nice ride, that last one. This engine is the smoothest one."

Several minutes later, the large display in Mission Control shifted from a map of the Moon to one showing Earth—lit up in red and green. A six-foot Christmas tree, twinkling with lights and tinsel, was moved to the front of the room, where everyone could take in its splendor and see its bright blue Earth-shaped ornament at the top.

But neither Borman, nor Slayton, nor anyone else took it for granted that Apollo 8 was home free, or anywhere close to it. The crew still had to travel 240,000 miles, make sure the guidance system worked, separate from the service module that had kept them alive, then survive reentry at record speeds. Even if all that worked, parachutes had to open, a landing site had to be hit, and the command module had to survive intact.

One hour after lighting the SPS engine for TEI, Apollo 8 was 3,225 miles away from the Moon and traveling at 4,125 miles per hour. That speed would decrease until lunar gravity gave way to Earth's gravity, and then Apollo 8 would begin falling faster and faster to its final destination.

At this moment, it began to dawn on Anders just how far from home he was. Valerie and his children were a quarter of a million miles away, a distance he could hardly process. He recalled a memory from when he was five years old. At the Shriners Circus, he saw a performer climb up the main circus tent pole and announce to the crowd that he intended to dive into a small bucket of water on the ground. The man flew from his perch, hit his mark, and survived. With Earth just a tiny marble in the vast ocean of space, Anders thought to himself: "My bucket is even smaller than that guy's from the circus."

Apollo 8 would require midcourse trajectory corrections on its way back to Earth, but those wouldn't come until later. In the meantime, there was room in the flight plan for the crew to sleep. None of them had managed more than about two hours over the last day.

"I hope it won't disappoint anybody too much," Borman told Mission Control about two hours after leaving the Moon, "but Jim is just in a daze, and so am I."

"Roger. No sweat," CapCom Mattingly answered.

Anders took the controls. A few minutes later, he put the ship into barbecue mode. He and his crewmates took a look back at the Moon. From this point forward, given the spacecraft's planned orientation, it was possible they would never see it again through their cabin windows.

Anders flew for three and a half hours before Borman and Lovell relieved him. CapCom Carr reminded Anders to hang up his Christmas stocking before falling asleep. The mood—on the spacecraft and in Houston—seemed relaxed, just right for the long coast home.

A few minutes later, Carr delivered the latest Interstellar Times report to Apollo 8. Los Angeles Dodgers pitcher Sandy Koufax had become engaged to his girlfriend, Anne Widmark, daughter of actor Richard Widmark. A Japanese exploratory party had shared a traditional Christmas dinner with Americans at the Navy's South Pole base. In California, liberated crewmen from the captured USS *Pueblo* donated their first paychecks to staff at the Naval hospital who'd cared for them on their return.

In Nevada, a little boy had written to Santa asking him to come in through the front door since his family had no chimney. "You will have to kick the bottom a little bit because it sticks," he warned. Near Palm Springs, California, a rabbi volunteered to serve as police chief for the day so the regular chief, a Methodist, could spend Christmas with his family.

In Moscow, a Soviet scientist predicted Apollo 8 would open the door to more cooperation between his country and the United States. In Cuba, Radio Havana rebroadcast the Voice of America program, allowing everyone there to learn of Apollo 8 and of the historic words spoken by the crew. Christmas shoppers in London crowded department stores and pubs to watch coverage of the lunar journey. The famed British astronomer Sir Bernard Lovell, who'd been critical of the mission and its myriad risks, expressed deep admiration for the flight. Pope Paul VI said the mission brought honor to these pioneers of mind and adventure.

Only a few naysayers popped up, most notably Samuel Shenton, founder of the England-based Flat Earth Society, who said the public was being hoodwinked by NASA.

"How does that grab you, Frank?" Carr asked during his report of the headlines.

"It doesn't look too flat from here, but I don't know, maybe something is wrong with our vision," Borman replied.

Carr finished by describing the scene at the astronauts' homes—the Christmas trees, wreaths, red bows, even the fake snow on the lawns. It was nearly 7:30 A.M. in Houston on Christmas morning, and all through the Borman, Lovell, and Anders homes, the children were awakening to celebrate.

An hour later, Susan Borman, her sons, and Frank's parents were in church. Susan was dressed in a powder blue boatneck dress with a white coat, and she held a thin, square cardboard box containing a reel-to-reel tape. She had the recording played for the congregation. It was a prayer for peace that Frank had read from space, and the verses from Genesis that had been read by the crew. Reporters had followed the Bormans to church, and Susan paused to speak with them on the way out. She told them she hadn't opened any of the presents under her tree, and she didn't intend to until Frank came home. "We'll be each other's big present," she said.

At the Lovell home, the children scrambled over each other to unwrap their gifts, many of which were delivered by a family friend wearing a Santa suit. When young Jeffrey ran outside to show the gathered reporters the toys he'd received, Marilyn noticed a photographer from the Associated Press, one of the nicer members of the press corps who'd been covering the family, standing in the cold. She walked outside to talk to him.

"Why don't you go home to your family?" she asked. "It's Christmas."

"I can't until I get a picture," he replied.

"Okay, wait a minute," Marilyn said, then went back into her house.

A few minutes later, the Lovell children came outside, each holding a new toy—a pogo stick, race cars, a yellow helicopter. The photographer snapped away. When he finally had his fill, Marilyn wished him a merry Christmas—and gently urged him on his way.

Although she was sure the man had gone, the doorbell rang again. This time the man standing before her was finely dressed and wearing a chauffeur's cap. Parked behind him was a Rolls-Royce. In his arms he held a box from the Neiman Marcus department store, beautifully wrapped in blue foil and decorated by two sequined spheres, one colored like the Earth, the other like the Moon. When Marilyn looked closer, she could see a little toy spaceship hovering over the lunar surface.

Opening the box, she moved aside tissue paper decorated with silver stars and found a mink jacket. The best part was the card. It read TO MARILYN, FROM THE MAN IN THE MOON. Marilyn put the jacket on over her pajamas and set about tidying the house, her feet hardly touching the ground as she twirled to dust shelves, glided to straighten pillows. Perhaps inspired by the emotions of the day, her thirteen-year-old son, Jay, kissed his mother for the first time since he'd been in grade school.

A few hours later the Lovells were in church. Although temperatures in Houston were heading into the midsixties, Marilyn wore her new mink jacket. Two-year-old Jeffrey, dressed smartly in a tan coat and a hat with a chinstrap, brought along his toy helicopter, which broke while he squirmed during the service. His sister helped calm him by walking him outside.

Valerie Anders also took her children to church. All her boys were dressed in suits, and her ten-year-old son, Glen, served as altarboy. During services, Valerie gave thanks for Bill's successful departure from the Moon and prayed for a safe return to Earth.

By this time, the astronauts' families had composed Christmas wishes for their husbands, which Carr delivered by radio to the crew 209,000 miles away. He spoke first to Lovell.

"Christmas morning around your house was kinda quiet, says Marilyn. She said that they're all thankful the mission has gone so great. They missed having you around the tree this morning, but they wanted to reassure you that your presents are waiting, and the roast beef and Yorkshire pudding will be on the table when you get home."

"Hey, that sounds good, Jerry. Good old roast beef and Yorkshire pudding," Lovell said.

"Hi, Frank," Carr said, now speaking to Borman. "Christmas morning has come at the Borman house. And the boys and Susan and your mom and dad all send their love. They say for you to stay in there and pitch."

These words meant a great deal to Borman. That had been his father's motto during the Depression, after he'd lost his gas station lease and things looked bleakest for the family, when he'd taken two jobs, changing tires and driving a laundry truck. In a more private time, Borman might even have cried thinking about all his dad meant to him. But here, on a mission, he remained a commander.

"Okay, thank you," he radioed back to Houston. "Please reciprocate for me."

Carr had a message for Anders, too, but Anders was sleeping and would get his later.

Apollo 8 coasted for another two and a half hours, its velocity dropping as lunar gravity continued to act on the ship. At a distance of about 39,000 miles from the Moon was the equigravisphere, the point at which Earth's gravity became dominant. Home was still 200,000 miles away, but now the spacecraft began to fall faster, a gradual acceleration that would take it to a speed in excess of 24,500 miles per hour at reentry into Earth's atmosphere. But that was a long way off, and for now, when the crew looked out their windows, with no landmarks in sight, they seemed to be standing still.

A little more than an hour after crossing the Earth-Moon gravitational divide, the astronauts began to prepare for the first of two television broadcasts scheduled for the return journey. Before showtime, CapCom Mike Collins settled some business. He started by delivering a Christmas message to Anders, who'd finally awoken.

"Valerie said to tell you that she and the kids are leaving for church about eleven thirty and eagerly awaiting your return. She said presents are magically starting to appear under the Christmas tree again, so it looks like a double-barreled Christmas."

"You can't beat a deal like that," Anders replied. "How was Christmas at your house today?"

"Early and busy as usual," Collins said. "I told Michael you guys are up there, and he said, 'Who's driving?'" Michael was Collins's five-year-old son.

"That's a good question," Anders replied. "I think Isaac Newton is doing most of the driving right now."

Collins informed the crew that Borman's family was at Mission Control. Susan did not want to distract her husband; instead, she and her boys, and Frank's parents, just beamed their grins to Collins.

"You've got a whole row of smiling faces in the back room, Frank," he said.

A few minutes later, the Borman family left to return home. Only then did the public affairs officer announce to the media that a message had been forwarded by NASA that morning to Mrs. Lloyd Bucher, wife of the captain of the USS *Pueblo*, whose crew had been held captive by North Korea. The message read, "You have been in our thoughts and prayers. Your reunion has brought great joy to our hearts this Christmas. Our best to you personally, and to all the families under your command. Signed, the families of the crew of Apollo 8." The note, the public affairs officer said, had been Susan Borman's idea, and she'd written it herself.

Shortly before going live on TV, Borman got ready to institute the planned midcourse correction. At an altitude of about 193,000 miles above Earth, he positioned the ship and fired its thrusters for fourteen seconds. In a short time, planners in Houston would know how much the burn had helped, but already they could see that the Trans Earth Injection maneuver used by the crew to leave lunar orbit had been nearly perfect.

Twenty-four minutes after the midcourse burn, Apollo 8 went live on the air. It was 3:15 P.M. Houston time, Christmas Day. The first pictures were of the spacecraft's complex instrument panel, then of Borman in his commander's seat. In front of him, a pair of legs floated upside down.

"Well, good afternoon," Borman said. "This is the Apollo 8 crew."

"It looks like you're okay, but somebody else is upside down," Collins said.

"Okay, that's right. That's Jim Lovell," Borman said, as if there was nothing unusual about that. "What we thought we'd do today was just show you a little bit about life inside Apollo 8. We've shown you the scenes of the Moon, scenes of the Earth, and we thought we'd invite you into our home."

Anders, working the camera, followed Lovell into the Lower Equipment Bay, where Lovell gave a demonstration of how the crew exercised (and bumped his head on navigation equipment, a detail Borman didn't fail to point out). Borman showed the command module's computer and its input keypad, then changed cameras to show Anders, who demonstrated how the crew ate in space, his meal floating before him.

"The food that we use is all dehydrated; it comes prepackaged in vacuum-sealed bags," Borman explained. "You notice that all Bill has to do to keep it in one place is let go of it. Except for the air currents in the spacecraft, it would stay perfectly still. He gets out his handy, dandy scissors and cuts the bag. The food is varied, generally pretty good. If that doesn't sound like a rousing endorsement, it isn't, but nevertheless, it's pretty good food."

Anders's dinner sounded appealing enough: corn chowder, chicken and gravy, sugar cookies, orange drink, and hot cocoa. He showed how to use scissors to open the freeze-dried orange beverage, inject water, and make it drinkable.

The men smiled and made jokes. Lovell showed off his navigation gear. Borman wished everyone back home a merry Christmas. Anders zoomed in for a close-up of Lovell's mission patch—the one showing a figure eight around the Earth and the Moon. Then the screen went blank. Even a few days ago, that design had seemed the fancy of a science fiction writer. Today, it had almost all come true.

After the broadcast, the crew broke out their own Christmas dinner. Each man expected more of what Anders had shown the world—dehydrated kibble—but that is not what they found. Wrapped in colorful holiday ribbons (fireproof, of course) and labeled with Merry Christmas messages, they uncovered a bounty: roast turkey with gravy so thick it didn't even levitate from the tray, stuffing, cranberry sauce. The topper was a gift smuggled in by Slayton himself—a two-ounce bottle of Coronet VSQ California brandy (100 proof) for each man.

"Put it back," Borman ordered when he saw the liquor.

If anything went wrong during the remainder of the flight, no matter how minor, the media would blame it on drinking, and there was no way Borman would risk that. Lovell just smiled. He and Anders had no intention of consuming the brandy; still, he thought Slayton's gesture was just about the greatest thing in the world. The men had just been to the Moon—*the Moon*—and a little romance was called for. No matter; it was enough just to look at the bottle's red and brown label, admire its fancy script ("Connoisseurs will delight," it promised), and dream of California, where the brandy had been awarded first prize at the state fair.

Each man's wife had sent along a Christmas package for her husband (NASA requirements: small, fireproof, under eight ounces), and now it was time for those to be opened. Lovell received cuff links and a tie tack with a moonstone; Anders a tie tack emblazoned with a gold numeral 8; and Borman a set of cuff links fashioned from a Saint Christopher medal worn by a family friend during battle in World War I. The men often dressed formally for official engagements, so the gifts would come in handy.

There was another set of cuff links on board. Twenty years earlier, as a plebe at the Naval Academy, Lovell had attended the Army-Navy football game in Philadelphia. There he met another plebe, just for a moment in passing, this one an Army man from West Point. The two strangers exchanged one of their two cuff links. It was only years later that Lovell learned that the man with whom he'd traded was Ed White, who'd also become an astronaut and who'd died in the Apollo 1 fire. To honor their friendship, Lovell had brought his own mismatched set of cuff links to the Moon.

The crew had a long stretch ahead of them, and the flight plan allowed for more downtime than it had for the outbound journey or the orbits at the Moon. That gave the men time to rest, and to reflect on the journey they had taken. To each of them, Earth still appeared tiny, just a far-off speck in an endless galaxy. To each of them, it seemed a miracle that all the events and conditions necessary for life had come together in just the right way at just the right time to create their home planet, and that they had gotten lucky enough to be part of it for just the briefest moment in the universe's still-unfolding story.

Chapter Twenty-Three

●

HELP FROM AN OLD FRIEND

BORMAN CLIMBED INTO HIS HAMMOCK FOR MUCH-NEEDED sleep, and Anders took control of the spacecraft. He found himself a bit bored; after discovering the Moon, even a swan dive to Earth could pale in comparison. Thousand-mile intervals ticked by as if counted off by metronome. All was steady.

While Anders controlled the spacecraft, Lovell took sextant sightings. Part of how Apollo 8 kept its attitude—the way it was oriented in space—was by aligning itself with the stars. To do that, Lovell would pick out known stars, then mark their positions through the onboard sextant. The computer would calculate the ship's attitude based on its relative position to those stars and automatically fire the ship's thrusters in order to keep its position in the desired orientation. It wasn't just important to be pointed toward Earth; it was important for Apollo 8 to be positioned the right way while it followed that path, especially upon reentry into the atmosphere, when its orientation had to be perfect.

Lovell had become a maestro at this job, "shooting" stars, entering data, and aiming the sextant like a concert pianist playing a Steinway. In fact, Lovell had earned the nickname Golden Fingers for his proficiency at punching these keys. But he was human, and not immune to a bad note. Early on December 25, Houston time, Lovell missed a step. He meant to enter Program 23 and then select Star 01. Instead, he entered Program 01 into his computer.

An alarm rang out. Suddenly, Apollo 8's guidance system reset itself, losing all memory of how the ship was oriented in space. As a result of Lovell's mistake, the guidance system now believed Apollo 8 to be back on the launchpad at Cape Kennedy. No one—not the crew, not the computer, not Houston—knew which way was up anymore.

Anders checked the eight ball—the attitude indicator that showed the spacecraft's orientation relative to the celestial sphere—and saw it moving and thumping in ways it shouldn't. At the same time, he heard one of the spacecraft's thrusters fire. Instantly, he was transported back to 1966, when he'd been on duty as CapCom for Gemini 8, the mission that had gone terribly wrong when a thruster could not be shut down, causing the ship to go into a violent and ever-accelerating tumble. Astronauts Neil Armstrong and Dave Scott had fought to stay conscious and regain control, which they finally managed by disabling the primary system and firing a set of reserve thrusters. As Anders saw the eight ball on his own spacecraft spin, he believed the same to be happening to Apollo 8.

To counteract the rotation, Anders used his hand-controlled thruster, but the ball just kept moving, so Anders added more thrust. Soon the spacecraft was in full rotation, and it was anyone's guess which way the nose was pointing in the cosmos.

By now, Borman had awoken from the commotion.

"What the hell's going on?" he called out.

"Stuck thruster!" Anders answered. But what to do? Armstrong and Scott had been in Earth orbit when it happened to them, and they barely made it out alive. Here, 185,000 miles farther away, it might be impossible to pull Apollo 8 out of its tumble and get it pointed back toward home.

Anders had stared down Soviet bombers, had landed on sheets of ice in his fighter planes, and always he'd run cool. This scared the hell out of him.

He looked back at the eight ball to assess the spin. Now it had frozen and was therefore useless. Yet the spacecraft was still turning—Anders knew it because he could see the cabin rotating against the pattern of sunlit dust motes floating freely inside it.

By now, Lovell had called to Anders and Borman that he'd made a mistake by resetting the guidance system. That explained why the eight ball had rotated to its launch orientation. And that explained why a thruster was not the problem. The thruster Anders had heard had fired automatically by program, a coincidence.

Furious at Lovell's mistake, Borman made his way to the control area, but already Anders had begun to fight the spacecraft's roll. He couldn't use the seized-up eight ball to judge how to rotate the ship, so he turned to a more ancient indicator the men had on board.

The dust.

Firing his thrusters, Anders turned Apollo 8 just enough to move the cabin back in the direction of the floating dust particles. When the cabin no longer moved in relation to its interior dust, he knew the spacecraft had stopped rotating. Apollo 8 was now in a steady attitude. But no one knew which way it was pointed, and the guidance system still said the ship was on the launchpad.

Borman took the controls. In an airplane, a pilot could eyeball the horizon and his runway and his surroundings if his instruments failed. Astronauts far from Earth couldn't do that. Borman and Anders now worried deeply about their ability to be sure of the spacecraft's attitude as they approached reentry; it was crucial that Apollo 8 be properly oriented by the time they hit the atmosphere in order to make a safe entry back into the world.

Everyone was angry: Borman and Anders at Lovell, Lovell at himself. It was a life-or-death situation. If they could not figure out how to reorient the spacecraft, Apollo 8 might not survive. The best idea, they agreed, was to use the stars. If they could pick out just a few they knew, they could begin rebuilding the computer's idea of their orientation.

Out the windows, the Sun shone against an all-black cosmos. Borman, Lovell, and Anders strained to locate stars, but crystals from the ship's evaporators, along with crystals from their own urine, followed the spacecraft, all of them masquerading as stars. Even when the astronauts thought they could distinguish between these tagalong crystals and a genuine star, it was impossible for them to identify the star, which was necessary for the computer to do its thinking.

It was then that Lovell found help from an old friend.

Looking through the spacecraft's optical system, he spotted the Moon. Then, locating Earth, he began to form a rough idea of the spacecraft's attitude with respect to the thirty-seven stars stored in the computer's database. Now, when he saw a star out the window, he could make an educated guess about its identity. If the crew still didn't know exactly how they were positioned in the heavens, at least they knew the neighborhood.

On the ground, Collins began to relay the procedures that would help the crew regain an accurate attitude reference for the ship. Mission Control was given access to a part of the computer's memory that had been corrupted by Lovell's mistake—in particular, a set of numbers that defined how the guidance platform ought to be aligned with the stars. Once Houston had uploaded the correct values, Lovell began a process of righting the ship, rotating it and sighting on stars, until, after about half an hour, the men understood again how their spacecraft stood in relation to the universe around them, and to home.

It was just after nine o'clock on Christmas night in Houston when Mission Control offered another gift to the astronauts, one they'd been planning for some time.

In place of the usual beeps and technical talk, NASA beamed up a recording of "Joy to the World" by Percy Faith and his Orchestra. Anders got a chill when it played; the music seemed to be coming from every direction, almost divine. Then he heard "O Holy Night" by the Norman Luboff Choir, but as the spacecraft turned in barbecue mode, the sound became warped. Anders had become so entranced by the music he'd forgotten to switch over between Apollo 8's various antennas. Tuned wrongly in space, even Christmas music could become eerie.

A few hours later, Anders reported that the crew had seen the Moon only once during its journey home. They were now 150,000 miles from Earth and 90,000 miles from the Moon. Five hours after that, they were 20,000 miles closer to home, and traveling about 3,800 miles per hour. By now, Apollo 8 had been in flight for more than five full days. Borman jokingly told Houston that they should "spread out one of those banners" in the Pacific Ocean splashdown area, and that Apollo 8 would try to "coast through it." Twenty-four hours remained until scheduled splashdown.

Around five o'clock Houston time on the afternoon of December 26, after almost a full day of uneventful cruising, Apollo 8 reached the halfway point between the Moon and Earth. It had been thirty-seven hours since they left the Moon, but if all went well, it would take just another twenty hours to reach the Pacific Ocean. The astronauts had one television broadcast left, when they would address a nation growing ever more nervous about their safe reentry into Earth's atmosphere.

Marilyn Lovell was nervous, too. She grabbed her two eldest children, Barbara and Jay, and whisked them to Mission Control, where they could all watch. As they arrived, they heard Jim poke a little fun at himself.

"I tried to hurry up the voyage home by calling up Program 01 to get us back on the pad, but it didn't work," he radioed to Houston.

"Well, that's the best excuse I've heard so far, Jim," Carr replied.

"The best of many," Lovell said.

A few minutes later, while millions of Americans watched, the crew of Apollo 8 began its sixth and final scheduled television broadcast. For nearly a minute, almost nothing appeared on-screen as Anders tried to frame the shot. But then a planet emerged, half lit, half in darkness, and there was no mistaking the swirls of clouds, the grooves of continents, the scoops of oceans. This was Earth, from 110,000 miles away. This, every person could see, was where they lived.

Lovell pointed to a storm over South America, the waters around the West Indies, and Florida. Looking through his telescope, he said he could see the central and southern United States. He asked Anders to describe his view.

"As I look down on the Earth here from so far out in space," Anders said, "I think I must have the feeling that the travelers in the old sailing ships used to have—going on a very long voyage away from home, and now we're headed back, and I have that feeling of being proud of the trip, but still . . . still happy to be going back home and back to our home port."

Nineteen hours remained until Apollo 8's scheduled reentry into Earth's atmosphere. With no major milestones due between now and then, the media was hungry for stories, and they turned to the astronauts' families to find them.

Valerie Anders reported that she was locked to her squawk

box. Her son, Alan, was playing with his dog, Luna, and cat, Dudley, while the other Anders kids concentrated on their Christmas presents. Valerie also noted that ten-year-old Glen had mowed the lawn that morning, a job Bill had given him before leaving.

Marilyn Lovell reported that she was trying to recover from all the excitement of the past several days. She told how daughter Susan had been jumping on her pogo stick.

Susan Borman said she'd spent much of the morning cleaning the house for friends who would join her for reentry and splashdown, and for Frank's return home. She noted that her two sons, seventeen-year-old Fred and fifteen-year-old Ed, would be helping her with clean-up duties. There was a reason for that.

Earlier that day, away from home, Fred and Ed had gotten into a fight, and Ed broke his left thumb throwing a punch to Fred's head. The boys knew this would be a problem and swore each other to secrecy. When they returned home, Ed walked around hiding his hand from Susan, but the pain only grew worse. The boys sneaked out of the house, drove to NASA, and found a doctor, who X-rayed Ed's hand. A short time later, Ed left wearing a giant white cast that reached halfway up his forearm. There would be no hiding that from their mother. When they returned home, Susan was angry that they could even think of fighting as their father plummeted toward Earth. At the same time, she was proud that they'd figured out how to handle the problem themselves. Helping her clean would be light penance.

About three hours after the television broadcast, Susan, Marilyn, and Valerie found rides to the home of astronaut Fred Haise, one of Apollo 8's backup crew, where a get-together hosted by his wife, Mary, was under way. These gatherings had become a tradition during space flights, and forty other astronaut wives welcomed the Apollo 8 ladies with ice cream and homemade cookies. Reporters dubbed the event a "hen party."

Several hours after the wives returned home, Apollo 8 was just 35,000 miles from the Pacific Ocean and had increased its speed to 7,700 miles per hour. By now, it was clear to Mission Control that the fourteen-second midcourse correction burn done a day ago had been so accurate that no further trajectory corrections would be needed. Backup recovery forces in the Atlantic and Indian Oceans were sent home. It would now be the Pacific Ocean or bust.

Four hours remained until scheduled splashdown. NASA's public affairs officer noted "unusually high" traffic in congratulatory messages flooding into the agency. *The New York Times* proclaimed, "There was more than narrow religious significance in the emotional high point of their fantastic odyssey, their reading of the biblical story of creation while this world watched live pictures of the Moon televised by the astronauts from within a few dozen miles of the lunar surface." Even acting NASA chief Thomas Paine couldn't contain himself. As the spacecraft drew closer to Earth, he wrote to President Johnson, "It is apparent that an unprecedented wave of popular enthusiasm for the Apollo 8 astronauts is building up around the world. Laudatory editorials are in every paper." Many were already calling the mission the greatest adventure in mankind's history.

But the ship wasn't home yet. Apollo 8 still had to hit the narrowest of entry corridors, at just the right attitude, moving at speeds faster than humans had ever traveled. To Borman and his crewmates, reentry was one of the three maneuvers on the mission—along with launch and Trans Earth Injection—during which they were most likely to die.

Reentry into Earth's atmosphere would officially start at an altitude of 400,000 feet, or 75.75 miles. Nothing magical happened at that point, but it's where things would start to change in a hurry. By that time, the command module would have shed the service module, leaving Apollo 8 just a cone-shaped wedge about eleven feet tall and thirteen feet wide speeding through space.

Several seconds later, Apollo 8 would have plunged 100,000 more feet, and Earth's atmosphere would begin acting on the ship and on the crew, exerting just a tiny fraction of a single g-force. The spacecraft would be traveling in excess of 24,500 miles per hour, and the computer would take over flying duties from Borman. At that point, the astronauts could only trust that Apollo 8 was aimed and positioned right. Some had compared NASA's challenge in finding the entry corridor to throwing a paper airplane into a public mailbox slot—from a distance of four miles. There was almost no margin for error.

If the spacecraft came in too steep, it would grind too hard into the atmosphere, causing massive g-forces that would crush the ship and crew, and generating heat so intense it would incinerate the men and turn Apollo 8 into a burning meteor.

If it came in too shallow, it would bounce off the atmosphere like a stone skipped on water and coast back out into space. Without the service module, Apollo 8 lacked any means of propulsion and could not apply the brakes sufficiently to reenter Earth's atmosphere. At that point, each astronaut would have a chat with his wife and children before drifting away from Earth in a ship with only a few hours' life support, to embark on a long elliptical orbit, one that would be fatal.

But even if the spacecraft hit the entry corridor perfectly, the friction created by the drag of the atmosphere on an object moving at almost 25,000 miles per hour would generate temperatures of 5,000 degrees Fahrenheit. To enable the astronauts to survive it, the command module had been covered by a heat shield made of a reinforced phenolic resin injected into a fiberglass honeycomb. Rather than defeat the heat in combat, the shield was designed to succumb to it and then vaporize away, leaving a new layer of shield beneath to continue the fight, all while keeping the command module cool. Even if it worked and the astronauts weren't fried to a crisp, they would be undergoing tremendous g-forces as the atmosphere slowed the ship. They would also lose all communications with Earth as gases around the spacecraft ionized from the shock wave, creating a kind of wall through which radio signals could not pass.

To mitigate the fantastic amount of heat and g-forces caused by reentry, Apollo 8 wouldn't simply plunge through the atmosphere; rather, it would use its aerodynamic design (its slightly off-center weight distribution turned the spacecraft into a kind of wing), allowing it to achieve lift and dip up and down, extend its path, shed velocity, and diffuse the heat that it had to endure as it aimed for the designated landing site.

The whole process would take about five minutes. If all went well, the spacecraft would have slowed enough to make its final drop to Earth. The astronauts' lives would then depend on the command module's parachute system, and the recovery forces that even now moved back and forth in the Pacific Ocean near Hawaii like predatory big cats on the hunt.

One hour remained until reentry. It was before dawn at the splashdown site, about a thousand miles southwest of Hawaii. It would still be dark when Apollo 8 arrived.

Traveling at 12,500 miles per hour about 11,000 miles above Earth, the astronauts stowed the last items still loose aboard the spacecraft. Given the huge g-forces of reentry and the jarring of the impact with the water, it was critical for the crew not to allow anything loose that could damage the cabin, or themselves.

Thirty minutes later, network television interrupted regular programming to cover reentry and splashdown. In Houston, it was just past 9:00 A.M. on December 27. All three astronauts' wives were watching their televisions at home, listening with one ear to the network anchor, the other to a squawk box. Inside the spacecraft, the crew cut off the oxygen flow from the service module, which they would soon leave behind. Their capsule now depended on its own small tanks for oxygen, a supply that wouldn't last much longer than the time required for a successful reentry and splashdown. Borman checked to confirm that the spacecraft was in its proper attitude.

Fifteen minutes before the planned reentry, Borman engaged a pyrotechnic sequence that severed cables and connections between the command and service modules, then blew apart the tendons that kept the two modules connected, a violent jolt that shook the astronauts. The service module had been Apollo 8's lifeline, housed its SPS engine, made every mile of this historic journey. Its built-in jets fired to thrust it away from the command module, lighting the sky in a final goodbye.

Only six minutes remained until reentry. Even now, the astronauts didn't wear their space suits or helmets, and wouldn't the rest of the way, a decision NASA had allowed so that the crew could equalize their eardrums manually as the pressure in the spacecraft rose during its descent through the atmosphere.

Inside the cabin, the crew was pointed backward and upside down, the windowed nose of the module facing back out toward space as they hurtled, still weightless and strapped in, at more than 20,000 miles per hour toward Earth's atmosphere. At launch, six days and two hours earlier, Apollo 8 weighed 6.2 million pounds. Now just 12,000 of those pounds remained. Looking out his window, Borman got a send-off from an old familiar face.

"Look who's coming there, would you?" he said.

"What?" Lovell asked.

"The Moon."

A minute later, Borman checked his indicators.

"Well, men, we're getting close."

"There's no turning back now," Anders said.

"Old Mother Earth has us," Lovell said.

Two minutes later, Anders noticed a change in the view outside his window.

"What is that?" he asked.

Borman and Lovell, the old spaceflight pros, decided to have a little fun with the rookie.

"That's right, you've never seen the airglow. Take a look at it," Lovell said.

"You can't get your [astronaut] pin without seeing the airglow," Borman said.

Apollo 8 plunged, blunt end first, toward Earth at more than 24,750 miles per hour, breaking below one hundred miles altitude and pushing its crew faster than any humans ever had traveled. On board, the astronauts, seated with their backs to the direction of travel, began to feel the first drag from the atmosphere and could see the dark sky begin to ionize and glow around them. Borman and Lovell might have experienced reentry before, but never at these speeds. They weren't kidding with Anders anymore.

"That's the airglow we are starting to get, that's what it is, gentlemen," Lovell said.

The three men braced themselves.

"Goddamn," Borman said. "This is going to be a real ride. Hang on!"

Chapter Twenty-Four

●

THE MEN WHO SAVED 1968

THE WORLD OUTSIDE THE SPACECRAFT LIT UP EVEN BEFORE THE astronauts expected it.

"I've never seen it this bright before!" Borman told his crew-mates. "You got zero point oh-five g yet?"

"Zero point oh-five g!" Lovell answered, checking a readout on the console.

"Okay, we got it!" Anders called.

The spacecraft neared 25,000 miles per hour.

"Hang on!" Borman yelled.

Out his window, Lovell could see a pink glow turning brighter by the second, and he felt the g-forces building. Temperatures rose fast around the command module as it collided with the atmosphere. The crew could only hope the heat shield would do its job; no manned ship had ever endured the heat loads Apollo 8 was about to experience.

A second later, Houston lost contact with the spacecraft as

Apollo 8 became enveloped in ionized gas. On CBS, Walter Cronkite narrated over an animated rendering of the command module entering a fiery atmosphere. At their homes, the astronauts' wives watched the broadcasts, willing their husbands home in these hand-drawn capsules.

Inside the spacecraft, the g-forces increased fast.

"They're building up!" Lovell called.

"Call out the g's," Borman told him.

"We're one g," Lovell answered.

The men's labored breathing could be heard on their intercom system as the forces multiplied.

"Ohhh!" Lovell groaned. "Five!" he called, straining to speak. "Six!"

Cronkite explained to the nation what the astronauts were enduring.

"Seven g's is seven times their weight on Earth, so these one-hundred-fifty-pound astronauts weigh something like one thousand fifty pounds, would be the effect as they are pressed against their couches."

Apollo 8 crashed even harder into the atmosphere. Despite the g-forces making it difficult to move, or even breathe, the ride was smoother than on lift-off, and the astronauts could still look out their windows and see the pink gases of the ionizing atmosphere turn a deeper reddish-blue; to Anders, he and his crewmates looked like flies caught in the middle of a blowtorch flame. In the distance, a Pan Am pilot flying in the darkness from Hawaii to Fiji watched the fireball created by Apollo 8 and estimated its cometlike tail to be more than one hundred miles long. Moments later, at maximum g-force load, the inferno surrounding the astronauts turned pure white as the temperature at the surface of the spacecraft rose to half that of the surface of the Sun. Out his window, Anders saw a terrifying sight—baseball-sized chunks of the heat shield flying off, many times larger than the grain-sized pieces NASA expected—and he waited for the heat to sear through the spacecraft and melt the crew.

But Apollo 8 did not melt. Instead, after about a minute of this peak intensity, the onboard computer automatically began to roll the spacecraft. Though the ship had no wings, its designed shape and offset center of mass made it capable of lift if positioned correctly, and now it began to climb a bit back out of the atmosphere, lowering the g-forces and cooling down in the process.

"Cabin temperature is still holding real good," Anders called to his crewmates, sounding a bit astonished and relieved. "Quite a ride, huh?"

"Damnedest thing I ever saw," Borman said.

Three minutes after losing contact with the spacecraft, Houston began calling to the ship, but CapCom Ken Mattingly couldn't get through. Apollo 8 had swooped back down for its second grind into the atmosphere. The crew held on.

"How much will this one go up, do you think?" Anders asked of the building g-forces.

"Three!" Lovell called.

Twenty seconds or so later, the spacecraft had been slowed by the atmosphere to suborbital speeds and began rolling one way, then the other, as the computer steered them toward the recovery ships. It had been nearly five minutes since the crew lost contact with Houston, but now Lovell began calling home.

"Houston, Apollo 8. Over."

Mattingly made out the voice through the static.

"Go ahead, Apollo 8. Read you broken and loud."

Borman jumped in.

"Roger. This is a real fireball."

One hundred thousand feet below, the USS *Yorktown* found Apollo 8 on its radar. A minute later, the spacecraft was at an altitude of just 40,000 feet and plummeting at a speed of about 680 miles per hour.

At around 30,000 feet, an altitude sensor fired explosives to jettison the top of the heat shield at the pointy end of the spacecraft. A moment later, two drogue parachutes shot out of the ship, making a giant *thwack* that Borman heard as they streaked up into the sky. The ship jolted when their lines went taut. These were not the chutes that would lower the craft to the water, but rather the smaller ones designed to stabilize Apollo 8, to keep it from wobbling and make it ready for the primary chutes. By the time they were out, the spacecraft was just 20,000 feet above the Pacific, but now its descent rate had slowed. Inside the cabin, an air vent opened to equalize inside and outside pressures.

Falling at a speed of 300 miles per hour, Apollo 8 rode gravity until an altitude of 10,000 feet, when the three main 80-foot parachutes were deployed. When their lines pulled tight, the spacecraft jerked hard. Anders worried that he'd felt only one jerk, not three. He knew that the Soviets had experienced trouble with their parachutes and that the technology, in general, was unreliable. Neither he nor the others could see the parachutes in the dark, but when Borman and Lovell checked their instruments, they could tell that the craft's sink rate had declined significantly, indicating proper functioning of all three chutes.

With the red-and-white parachutes fully blossomed, Apollo 8's descent rate fell to just 19 miles per hour. On board, thrusters were ignited and their tanks purged of propellant to prevent harmful substances from polluting the splashdown and recovery area. The fire spitting from the burning thrusters lit the still-dark sky, giving the astronauts their first view of their parachutes, and good reason to believe they were floating down as planned. Under the chutes' risers, the capsule was tipped on an angle to allow it to knife into the water rather than belly-flop onto its blunt base. At an altitude of just 8,000 feet, Apollo 8 was less than five minutes from scheduled impact with the water.

Moments later, one of the recovery aircraft made radio contact with the spacecraft.

"Welcome home, gentlemen," a crewman called to the astronauts, "and we'll have you aboard in no time."

At three minutes to splashdown, recovery helicopters spotted flashing beacons from the falling spacecraft. Apollo 8 was almost directly over the *Yorktown,* a bull's-eye of almost unimaginable accuracy.

"Stand by for Earth landing!" Borman called from his commander's seat.

At their homes in Houston, the astronauts' wives stared at their televisions. For Valerie, it was thrilling to hear that the parachutes had opened—that meant Apollo 8 was somehow reconnected to Earth. But she thought, "They're heading for a big, dark, rough ocean, and the ships still don't know where they are."

At one thousand feet altitude, radio traffic from the recovery forces grew so voluminous that the astronauts couldn't communicate with one another.

"Turn him down!" Anders told his partners. "Christ, we can't get anything done."

Just a hundred or so feet remained. The crew braced themselves, not knowing exactly when impact might come.

Borman called to Lovell and Anders.

"Maybe we better get these—"

At that moment, Apollo 8 came in flat, not on its intended angled edge, and bashed into the Pacific Ocean, its blunt end colliding against the upswell of a wave, just about the most violent impact possible. Inundated by water (and perhaps stunned by the crash), Borman could not flip the switch to cut the parachutes from the capsule, and Apollo 8 was dragged over by its chutes and turned upside down in the ocean. None of the men was ready for an impact that jarring; nothing in simulation had come close. By the time Borman came around a few seconds later and cut the lines, all three men were hanging upside-down in their straps. Garbage that had collected in the cabin streamed down on them, and water poured over their bodies and faces.

Right away, the crew believed the spacecraft had split open from the impact and was flooding with seawater. Anders got ready to pounce on the hatch and open it, then get his crewmates and himself out before the ship sank—they'd trained for that kind of emergency—but a moment later he could see that no more water was running in, and he realized that the crew had been doused not by seawater but by condensation around the various cold parts of the spacecraft's interior. Anders could only smile at the picture: three conquering heroes returned from the Moon, hanging upside down and dripping in garbage.

Borman reached for a button and inflated three large balloons, which flipped the spacecraft back over, blunt side down. The men were now right side up in their seats, but it was too late. Sickened by the impact, the high seas, and the sudden inversion, Borman vomited all over his crewmates. It had been bad enough on the outbound journey when Borman threw up, but now his crewmates let him have it.

"Typical Army guy!" the two Navy men yelled at their commander. "Can't handle the water!"

Television cameras showed live images from the recovery ship and one of the rescue helicopters. Cronkite removed his glasses, as if he couldn't quite believe the journey had ended.

"The spacecraft, Apollo 8, is back, and what a remarkable trip and remarkable conclusion," he told the nation. "The spacecraft has landed within two and three quarters miles of the carrier . . . Apollo 8 has ended up to this point as perfectly as it began."

At home, the astronauts' wives were overcome by joy, relief, and wonder. Their children hugged their crying mothers. At Mission Control, applause broke out, and a fifteen-foot-long American flag was unfurled, one that eclipsed the giant wall map that had been used for the mission. All three flight shifts were present to experience the moment. "The Star-Spangled Banner" played in everyone's headsets.

"It is a veritable roar in here," the public affairs officer announced. "The room is awash with cigar smoke. A number of congratulatory messages are coming across this console . . . I've never seen a degree of this emotional outpouring in any previous mission, including Alan Shepard's . . . I've seen rallies in locker rooms after championship games, happy politicians after elections, but never, none of them do justice to the spirit pervading this room."

Some of the controllers and personnel had also brought along triangular flags with a white numeral 1 sewn in, to indicate victory over the Soviets. Someone at NASA, however, suggested that that might not be the most magnanimous of displays, and the men agreed. Instead, they waved American flags, which to many of them said it all.

In Rome, Pope Paul VI, who'd watched the splashdown on television, knelt in a prayer of gratitude. In Communist Cuba, state news covered the return of the spacecraft. World leaders began writing notes of congratulations to America.

In the capsule, Borman, Lovell, and Anders were still covered in vomit and garbage. The spacecraft had come through unimaginable heat with almost no effect on cabin temperatures, but now, bobbing on the waves, it began to grow hot inside, likely from retained heat sizzling upon impact with the water. Temperatures soon subsided, however, and Anders began to appreciate how beautifully the heat shield had worked. The huge chunks he'd seen flying off during reentry had really been just granular in size; surrounded by an ionized haze, and streaking by at thousands of miles per hour, they had appeared through his window like fiery baseballs.

The crew worked to remove their straps while helicopters circled above. Recovery forces itched to get to work, but NASA protocol required them to wait for the break of dawn and the onset of natural light, in about forty-five minutes, so all they could do was hover, close in, and shine lights around the bobbing capsule. Men armed with rifles scanned the waters to make sure no sharks were in the area during recovery time.

Just before first light, several swimmers dropped from their chopper into the water. When they reached Apollo 8, they affixed an inflatable collar to the spacecraft, stabilizing it and providing a platform on which to step and work. Through a window, one of the swimmers flashed a thumbs-up to Anders, who returned the gesture. While the *Yorktown* moved toward the recovery scene, one of the helicopter pilots radioed the astronauts with a question.

"Is the Moon really made of green cheese?"

"No," Anders replied. "It's made of American cheese."

Soon after, the *Yorktown* called to the capsule asking what the astronauts might like for breakfast. The answer was unanimous: biscuits, steak, and eggs.

As daylight broke, three swimmers worked to open the spacecraft's hatch. When it lifted, one of the swimmers stuck his head inside, only to recoil as if repelled by a force field. He soon found his feet and, along with the others, helped the crew of Apollo 8 out of the capsule. As they stepped onto the inflatable platform around the spacecraft, none of the three astronauts could imagine a smell sweeter than the fresh sea air—a smell they'd known forever and yet was new to them today.

A helicopter dropped a life raft into the water, and one by one, the astronauts climbed inside. The helicopter then lowered a basket-shaped net for the crew; one at a time, they were hoisted into the chopper. Anders was the last to go. Looking up at the helicopter, it struck him that almost everything on Apollo 8 had been designed with great redundancy, yet here he was, at the very end of his journey, hanging over the ocean by a single wire.

It was 11:14 A.M. Houston time when the helicopter closed its door. Looking back down toward his spacecraft, Borman gave thanks to the scalded machine, an exquisite piece of design and daring. A moment later, the chopper dipped its shoulder into the yellow-pink new sky and headed for the *Yorktown*. On the carrier, hundreds of crew dressed in Navy whites jammed the decks, eager for a glimpse of the returning pioneers.

On board the helicopter, a crewman handed Borman an electric razor. NASA had figured out how to get three men to the Moon and back again but still hadn't perfected technology that would allow men to shave without polluting the command module with stubble. When Borman had asked to arrive at the aircraft carrier clean-shaven, a portable electric razor on board the chopper was the best NASA could offer. Soon, Borman was whisker free.

On television sets at their homes, Susan, Marilyn, and Valerie watched as the helicopter slowed to a hover over the deck of the *Yorktown* and then set down. Ship's crew ran out, ducking their heads, to secure the chopper to the deck. After the rotors stopped, a short stair platform was brought to the aircraft door, and a red carpet was unrolled at the foot of the stairs.

The door opened and the three astronauts stepped forward, first Borman, then Lovell, then Anders. They smiled and waved, overwhelmed by the roar of the hundreds of sailors on board the ship, each of whom was away from home for Christmas, just as they were. A giant American flag held by the Navy color guard danced in the ocean breeze. In the sound and the moment, Borman's mind traveled back in time, over all the training and planning that had been done for Apollo 8, over the thousands of people who had worked so hard for this audacious mission, and he thought about how so few of them would ever be recognized like this, and how so many of them deserved to be.

Anders lost his balance for a moment when the astronauts finally made their way down the stairs, not an unexpected result after more than six days of weightlessness. Watching at home, Valerie thought that her husband looked skinny. It was the smiles that convinced the women that their husbands really were home safe. By now, Marilyn was so spent from the stress of the past week that she had little voice left and even less energy, but she couldn't remember ever feeling so happy. "He's beautiful," she did manage to tell reporters. "I know that's no way to describe a man, but he looks just beautiful."

At a nearby microphone, the *Yorktown*'s commander, Captain John Fifield, welcomed and congratulated the men. Taking the microphone, Borman addressed the ship. Millions of people watched around the world.

"We're just very happy to be here and appreciate all your efforts, and I know you had to stay out here over Christmas and that made it tough . . . We can't tell you how much we really appreciate you being here, and how proud it is for us to participate in this event, because thousands of people made this possible, and I guess we're all just part of the group. Thank you very much."

Surrounded by sailors, the astronauts made their way across the flight deck, then down an elevator to the hangar deck and into the ship's sick bay for a medical evaluation. For his part, Anders was in no shape for an inspection. As part of his plan to avoid defecating in space, he'd asked NASA doctors to prescribe a low-residue diet before and during the flight, and his plan had worked so well that he hadn't had a bowel movement during the entire mission. Now he needed to find a toilet.

He located a cabin just in time. As nature began to have its say, there was a pounding on the bathroom door.

"Major Anders! Quick! You've got to come to flag bridge. The president is going to call in five minutes. Move it!"

Anders was torn between his duties. He could only answer to the higher power.

"I'm not going!" Anders yelled to the man. "Tell him I'm on the toilet and I'm not going."

There was no way Anders could risk losing control of himself while talking to the president of the United States. A minute later, one of the ship's doctors ran in with a portable telephone and passed it through the bathroom door to Anders. Borman and Lovell picked up their own extensions, likely in sick bay, surrounded by physicians, stethoscopes, and syringes.

Less than a month remained in Johnson's presidency. Five years earlier, he'd taken over from his slain predecessor, a president who'd made an impossible promise: to land a man on the Moon by the end of the decade. Johnson might have been forgiven for backing off Kennedy's commitment. Instead, he'd charged forward.

"You've seen what man has really never seen before," Johnson said to the astronauts. "You've taken all of us all over the world and into a new era. And my thoughts this morning went back to more than ten years ago . . . when we saw Sputnik racing through the skies, and we realized that America had a big job ahead of it. It gave me so much pleasure to know that you men have done a large part of that job."

And it gave Borman, especially, the same kind of pleasure. He'd gone to the Moon because of his love of country, and because he felt it was important to beat the Soviets in the race to get there. He'd always told himself his mission wouldn't be done until he and his crewmates were standing on the carrier deck. Putting down the telephone after speaking with President Johnson, he knew he'd made it.

After the call, seventeen doctors, researchers, and medical technicians inspected the astronauts, taking blood, conducting tests, making sure all was well. Even a psychiatrist got his turn, looking for signs that such profound separation from home and family and Earth might have disturbed the men's psyches. Other than some stiff legs—and Lovell's lingering tendency to let go of things in midair and expect them to float—everyone checked out fine. Following their medical examinations, the crew were allowed to phone their wives; even from a distance of several thousand miles and through the thick static, these women never sounded so close.

The astronauts made their way back to the flight deck to thank the crew of the *Yorktown* and to meet with the swimmers who'd made the recovery. While shaking hands, Anders recognized the man who'd first opened the hatch of the spacecraft.

"That was really great, Corporal," Anders said. "I noticed, though, that when you poked your head in you fell backward. Was it the way we looked?"

"No, sir," the man replied. "It was the way you smelled."

The astronauts had a good laugh about that one.

By now, it had been several hours since splashdown. In Houston, Susan, Marilyn, and Valerie tried to adjust to the idea that they needn't worry anymore, that today was now just a regular Friday. At Mission Control, it was finally time to celebrate. Consoles were unplugged and data secured, and many of the controllers and managers returned to the haunts that had been bridges for the endless nights and years they'd spent working to get to the Moon. Some went to the Singing Wheel, some to the Flintlock, others to the Holiday Inn across from NASA in Houston. Most everyone drank and smoked cigars and raised toasts. At the Flintlock, John Aaron and Rod Loe, who'd worked with Anders to write mission rules and procedures, stood at the bottom of the stairs, not yet ready to go up and join the party.

"What are you guys doing?" a friend asked. "Why aren't you upstairs?"

Loe thought it over for a moment.

"We're just standing here thinking how proud we are to be Americans," he said.

Borman, Lovell, and Anders dined on lobster tails and roast beef that evening with Captain Fifield, then collapsed in comfortable beds made up with crisp, fresh sheets, getting their first good sleep in more than a week. The next morning, they enjoyed steak and eggs with some of the *Yorktown*'s officers.

That day, December 28, the astronauts boarded a carrier plane and flew from the *Yorktown* to Hickam Air Force Base in Hawaii. From there, they transferred to a C-141 transport plane for a flight of more than eight hours to Houston. For Anders, it would be the longest flight he'd ever endured other than the one aboard Apollo 8.

The plane reached Ellington Air Force Base after 2:00 A.M. on Sunday, December 29. Hundreds from NASA, and three thousand well-wishers, many holding banners with congratulatory messages, were there to greet the astronauts, who were clean-shaven, dressed in blue coveralls, and wearing baseball caps. Under a half Moon, Borman, Lovell, and Anders found their wives and children, gave them red and purples leis from Hawaii, and pulled them close. Eight-year-old Gayle Anders gazed at her father, grateful to have him back and not sure she should ever let go. The Borman boys wore ties and beamed at their dad. The Lovell kids orbited their father, pushing close for his attention, never staying on his far side for too long.

Borman stepped up to a microphone, his wife's red lipstick smudged across his face. "Thank you for coming out so early in the morning to welcome us," he said. Lovell added, "At two in the morning, I expected to get in my old blue bomb and go home." (Lovell's "blue bomb" was the family's no-frills 1962 Chevy Biscayne.) The astronauts thanked the crowd and their families at a microphone, greeted NASA's managers and controllers, and smiled for photographers. One boy in the crowd told his friend, "I know they didn't have radiation because I just shook their hands." Then it was time for the astronauts to go home.

But that wasn't proving so easy. The crowd pushed forward, surrounding the crewmen and their families, thrusting dollar bills to be autographed. In the surge, Bill Anders became separated from Valerie; NASA staff scurried to reunite the couple, but no one seemed to mind such a short separation, least of all the two of them.

Each of the families finally climbed into their car and drove off. In their rearview mirrors, the astronauts could see throngs of people waving goodbye until they'd pulled out onto the Houston roads and there was only black night behind them.

None of the men said much about his trip as he drove home. They just said how happy they were to be back, that all had gone as perfectly as could be imagined, and that they felt lucky. None of them was inclined to philosophize about the trip—not yet, anyway. Over the years, these men had become expert at coming home from missions, forgetting about the risks they'd just undertaken, getting on with their day. No other kind of men could have climbed into such unproven flying machines. These were the kind of men NASA had always wanted.

When Borman, Lovell, and Anders opened their front doors, they found Christmas trees still glowing and presents waiting for them, and they knew that this was just how their homes had looked on Christmas Eve when they had been 240,000 miles away at the Moon, and they knew that this was how their homes would have looked no matter how long it might have taken them to return.

The next morning, as the Bormans sat down for breakfast at the kitchen table, Frank asked the boys about football and hunting, and demanded to know why dog food had been left in the bowl while he was gone. As for Edwin's broken thumb—by the look on their dad's face, they knew there had better be a good explanation for that. When the family opened presents, Susan found a new dress Frank had bought for her before he'd left for the Moon. He'd always loved shopping with her, and knew her style and size.

In the days that followed, it seemed the world talked only about Apollo 8. A *New York Times* editorial called it "the most fantastic voyage of all times." The Washington *Evening Star* announced that "Man's horizon now reaches to infinity." The *Los Angeles Times* said the mission "boggles the mind." And *Time* magazine rushed to change its iconic Man of the Year cover from THE DISSENTER to ASTRONAUTS ANDERS, BORMAN, AND LOVELL.

Even the Soviet Union could not hide its admiration. Apollo 8, the nation said, "goes beyond the limits of a national achievement and marks a stage in the development of the universal culture of Earthmen." In a congratulatory note, several Soviet cosmonauts lauded their counterparts for "the precision of your joint work and your courage."

Telegrams for the astronauts poured in by the thousands. One, however, stood out from the rest. It came not from a world leader or celebrity or other luminary, but from an anonymous stranger.

It had traveled over whites-only lunch counters in the South, through jungles in Vietnam where young men fell, over the coffins of two of the America's great civil rights leaders. It had blown across streets bloodied by protesters and police, past a segregationist presidential campaign, into radios playing songs of alienation and revolt. It had made its way through ten million American souls who didn't have enough to eat, alongside generations that no longer trusted each other, into a White House where a no-longer-loved president slept.

It read:

THANKS. YOU SAVED 1968.

Epilogue

●

As the world celebrated Apollo 8, most didn't realize just how successful the flight had been. By NASA's analysis, all mission objectives had been attained. The command and service modules had performed beautifully at the Moon. Deep space communications had been excellent. Mascons—the anomalies in lunar gravity—were better understood. Navigation over lunar distances was proved with exquisite accuracy. Lunar landmarks were confirmed for future missions. And the Saturn V rocket, which had been so troubled on only its second test, performed almost flawlessly on its third.

Despite the breakneck pace at which they had been working since August, few at NASA took time off during the last hours of 1968, especially those responsible for analyzing photographs and movies returned by Apollo 8. Experts developed film by hand rather than by machine, a painstaking process that assured the film could be salvaged if mistakes were made.

Anders was at home in Houston on December 29 when some of the pictures were developed—pictures unlike any mankind had seen before. Shots he took of Earthrise showed the bright blue-and-white marble of Earth rising over the Moon's gray horizon, the only color in an all-black universe—a tiny, shining oasis in the cosmos.

NASA selected the best one of Anders's Earthrise photos, and on December 30, it appeared on the front pages of newspapers across the globe. Days later, it would run in full color in magazines and Sunday supplements. (Most everyone published the photo with the Moon's surface horizontal, though Anders and his crewmates had witnessed Earthrise with the lunar surface both horizontal and vertical. To Anders, both perspectives were correct—there was no real up or down in space.) In the following year, the United States Postal Service would issue a new stamp featuring the Earthrise image. In a year of historic photographs—of the street execution of a Vietnamese prisoner; of Martin Luther King, Jr.'s associates pointing in the direction of his assassin; of Robert F. Kennedy lying mortally wounded in a busboy's arms; of the Black Power salute at the Summer Olympics—the image of Earthrise captured the world's imagination.

In case anyone had missed it, President Johnson sent a print of Earthrise to every world leader. Grateful for the attention the photograph produced, the Hasselblad company, which made the cameras used by the astronauts aboard Apollo 8, offered a brand-new one to Anders. A stickler for regulations, he declined; by his reading, government rules made it illegal to accept gifts.

As America rang in the new year, NASA continued a series of debriefings with the crew of Apollo 8. Borman, Lovell, and Anders were together for these meetings with agency officials, where they provided comments on every aspect of Apollo 8 and made recommendations for future missions. Some remarks were matter-of-fact, as when Anders suggested better light control settings on television cameras. Others reflected excitement, as when Lovell said, "There are a tremendous amount of craters that are not picked up in Earth-based or Earth-orbital-based photography. There are many more new craters to be seen in lunar orbit," or when Borman described reentry: "The whole spacecraft was lit up in an eerie iridescent light very similar to what you'd see in a science fiction movie. I remember looking over at Jim and Bill once and they were sheathed in a white glow. It was really fantastic." As fascinating as these descriptions were, the agency wanted to complete their investigations into the flight as soon as possible, as the astronauts had a busy January on their hands.

It began at the White House on January 9, where the astronauts each received NASA's Distinguished Service Medal from President Johnson. From there, Borman, Lovell, and Anders rode in their motorcade through flag-waving throngs to Capitol Hill, where they provided an informal briefing to a joint session of Congress and the Supreme Court (and received a two-and-a-half-minute standing ovation). Lovell told the distinguished audience that a few days after returning from the flight, he'd walked outside his home in Houston and gazed up at the Moon. "I could scarcely believe I was there," he said. The men then moved to the State Department, where they held a press conference. Before opening the floor to questions, a NASA spokesman announced that Borman had been named deputy director of Flight Crew Operations, an administrative position that would ultimately involve advising the White House on NASA affairs.

One reporter asked: "Was there any moment during the mission in which you were a little bit scared or frightened?"

"I was scared or frightened during a lot of the phases of the mission," Anders said. "But I think that fear is a normal human reaction and one that is not detrimental to the flight as long as you keep it under control."

Another reporter asked, directly, when America would land on the Moon.

"This summer," Borman said.

"Can you be more precise?" the reporter asked.

"Apollo 11," Borman said.

What Borman didn't say was that he, Lovell, and Anders might have been the crew for Apollo 11—if only Borman had wanted it. Deke Slayton, who assigned crews, thought that the Apollo 8 astronauts were best positioned to train for the first Moon landing, since they'd already made a lunar orbit flight. But given Borman's decision that Apollo 8 would be his last trip in space, Slayton didn't need to further consider Borman and his crew. In the end, Slayton decided to stay with the planned rotation, with Neil Armstrong as commander of Apollo 11.

The day after the press conference, the astronauts were honored with a ticker tape parade in New York City, an appearance at the United Nations, and a party at the Waldorf Astoria with Mayor Lindsay and Governor Rockefeller. Celebrations and parades followed that week in Newark, Miami (at the Super Bowl), Houston, and Chicago, where more than a million people turned out.

During the festivities, the astronauts never forgot that there was a war going on in Vietnam in which their friends and colleagues continued to risk their lives and die for the United States. To Borman, Lovell, and Anders, countless men were doing more for their country in Vietnam, and elsewhere, than the astronauts had done by flying Apollo 8, men that no one would ever know about, brothers the crew of Apollo 8 tried to remember every day.

On January 20, following Richard Nixon's inauguration, NASA announced that Lovell and Anders would be on the backup crew for Apollo 11, the mission expected to make the first lunar landing. That meant if the primary crew—Neil Armstrong, Buzz Aldrin, and Mike Collins—made the trip, Lovell and Anders likely would be primary crew for another lunar landing, Apollo 14. However, Slayton had made Anders the backup command module pilot for Apollo 11, which meant Anders would be the one who stayed with the orbiting spacecraft while his two crewmates flew the lunar module to the Moon's surface and back. To Anders, the writing was on the wall just as he expected; he'd mastered the spacecraft so beautifully on Apollo 8 that no one would ever let him leave it.

In late January, President Nixon sent Borman and his family on an eight-country goodwill tour of Europe. At each stop, Borman was greeted as a hero, and people listened with rapt attention as he described his adventure around the Moon. In Rome, he and his family met the Pope and stood on the same piece of ground where Galileo had been condemned for heresy in 1633 for advocating Copernicus's argument that the Earth travels around the Sun. Of the crew's reading from Genesis, Pope Paul VI said, "For that particular moment of time, the world was at peace."

Flush with confidence and shot full of momentum from the success of Apollo 8, NASA entered its final push to land men on the Moon. It would require two flights—Apollo 9 and Apollo 10—to prepare for a landing mission sometime in the summer of 1969, but after Apollo 8, the space agency believed there was virtually nothing it couldn't do.

It was around this time that NASA asked Borman to talk about the space program and Apollo 8 at American colleges. Some welcomed him. Many more did not. Often, he was shouted down by protesters who resented the presence of a military man on campus. At Columbia University in New York, he was pelted by marshmallows, then overrun onstage by students dressed in gorilla costumes. In Boston, a helicopter had to deliver him past the mobs that blocked access to his speech. But the worst experience came at Cornell University, where astronomy professor Carl Sagan invited Borman and Susan to his home for a roundtable with students. The Bormans accepted, then were treated to an evening of attacks on America and its conduct in Vietnam, all of it encouraged by Sagan, whom Borman would never forgive for the treatment.

After Apollo 9 and Apollo 10 flew successful missions in March and May 1969, Borman and his family boarded a plane for another goodwill trip, this one to Russia. The Soviet ambassador to Washington had extended the invitation, and President Nixon thought it a positive step toward easing tensions between the two nations. With the president's permission, Borman and Susan brought their sons. As the first astronaut ever to visit Russia, Borman was given first-class treatment and shown some of the Soviet Union's proudest sites. He was given a tour of the highly secret "Star City" near Moscow, where cosmonauts lived and trained; laid wreaths at the resting places of Soviet leader Vladimir Lenin and cosmonaut Yuri Gagarin; sampled wines in Yalta; and met with a top Soviet physicist. One night, he and his family were honored at a crowded dinner at the regal restaurant in Moscow's famed Metropole Hotel. One of the cosmonauts in attendance, Alexei Leonov, was struck by a piece of Borman's attire—in place of a traditional necktie, he wore a bolo tie set with a bright blue stone. "Everyone wanted to stand near to him," Leonov would later write of the evening. "To touch him."

Borman congratulated Leonov on his 1965 spacewalk and described how the Moon had appeared close up, then showed slides of his lunar journey to a rapt audience. Although Borman had considered the Soviet Union an enemy, he liked the Russian people and held the cosmonauts in the highest regard—no westerner better understood the rigors of their training, or the great risks they took for their country. Susan took a ring from her finger and gave it to a cosmonaut, a gift from her family to theirs. By trip's end, Borman saw the cosmonauts as he saw the astronauts—a group of test and fighter pilots, all of whom wanted more than anything else to help their country succeed. And he admired their candor—to a man, they seemed generous in acknowledging that America had won the race to the Moon.

On July 16, 1969, NASA launched Apollo 11 from Cape Kennedy. On board were astronauts Armstrong, Aldrin, and Collins. Orbiting the Moon, Armstrong gave a shout-out to Apollo 8.

"We're over Mount Marilyn at the present time," he radioed to Houston.

A day later, July 20, Apollo 11's lunar module set down on the lunar surface, at one of the sites at the Sea of Tranquillity scouted by Apollo 8. "The Eagle has landed," Armstrong radioed to Mission Control. Six and a half hours later, Armstrong exited the spacecraft, climbed down its ladder, and set foot on the Moon.

"That's one small step for [a] man; one giant leap for mankind," Armstrong told the world.

Aldrin followed onto the lunar surface several minutes later.

Just eight years after a young president had pledged to do the impossible by decade's end, America had made good on his promise. Men had landed on the Moon.

In early August 1969, the atheist Madalyn Murray O'Hair and her affiliated group brought a lawsuit against NASA and its chief, Thomas Paine, for allowing the reading of the first lines of Genesis by the crew of Apollo 8 during lunar orbit. That action, O'Hair claimed, abridged her First Amendment right to be free from religion and violated the First Amendment by establishing Christianity as a state religion. Further, she alleged that NASA had chosen Christmas for Apollo 8's mission for religious reasons. She asked the court to prevent NASA, a public agency, from future religious displays or readings. A United States district court judge threw out the lawsuit; the United States Supreme Court refused to hear O'Hair's appeal.

By the time Apollo 11 had returned from its historic mission, some shuffling had gone on with Apollo crews. Anders, who believed he'd never be given the chance to walk on the Moon, accepted a job at the National Aeronautics and Space Council, in Washington, D.C., a body chaired by the vice president of the United States and devoted to establishing the country's space policy. (Anders took the job on one condition: that he retain his astronaut status in case a miracle occurred and Deke Slayton gave him a walk on the Moon.) Lovell had been advanced in the cycle to command Apollo 13. For his part, Borman had become a special adviser to Eastern Airlines.

Following a successful landing mission by Apollo 12 in November 1969, Apollo 13 launched, with Lovell as its commander, on April 11, 1970. That flight, however, never made it to the lunar surface. Near the Moon, an oxygen tank exploded, severely damaging the service module and disabling the command module's power and oxygen supply. Two hundred thousand miles from Earth, and unable to safely fire the SPS engine, Lovell and his crewmates, Jack Swigert and Fred Haise, were in grave danger of being stranded in space forever.

Tapping wells of ingenuity, creativity, and courage, Mission Control and the crew cobbled together a solution. They would use the lunar module for electrical power and thrust, repurposing it to keep themselves alive and to regain a free-return trajectory that would whip them around the Moon and return them to Earth. The world prayed for the astronauts as they sped home in their freezing spacecraft, while Lovell used the experience he gained from restoring Apollo 8's disrupted orientation during its own return to nurse the disabled Apollo 13 back to Earth. No one knew whether the command module's heat shield had been too severely damaged by the explosion to survive reentry.

In the end, Apollo 13 made it back safely, one of the great rescues in history. No one at NASA, least of all Lovell, failed to recognize that the crew had been saved by the lunar module's secondary role as a lifeboat. It was this lack of a lifeboat that had haunted so many who'd feared flying Apollo 8 to the Moon. If the explosion aboard Apollo 13 had occurred during Apollo 8, Borman, Lovell, and Anders would never have come home.

Five days after Apollo 13's return, the first Earth Day observance was held, a series of demonstrations, celebrations, and rallies to protect the environment. Apollo 8's Earthrise photo was used as the movement's symbol. Some suggested it was Apollo 8 itself—man's first look at his home planet, and at its thin, fragile atmosphere—that launched the environmental movement.

NASA made four more manned trips to the Moon after Apollo 13, all of which successfully landed crews on the surface. Collectively, the astronauts on the Apollo missions returned almost 842 pounds of lunar soil and rock, samples that continue to form the bedrock upon which our understanding of the solar system's origins is based. In all, twelve Americans walked on the Moon between 1969 and 1972.

And that was it.

Since Apollo 17, humankind has never returned.

It has been fifty years since Apollo 8 flew to the Moon.

Many of the key managers at NASA who made the mission happen have since died, including former administrators James Webb and Thomas Paine; Director of NASA's Manned Spacecraft Center Robert Gilruth; Chief of Astronaut Division Deke Slayton; Associate Administrator for Manned Space Flight George Mueller; Director of NASA's Apollo Manned Lunar Landing Program Samuel Phillips; and rocket mastermind Wernher von Braun.

All three astronauts who served as CapComs during Apollo 8—Mike Collins, Ken Mattingly, and Jerry Carr—are still living. Collins went to the Moon as part of Apollo 11, Mattingly on Apollo 16, both as command module pilots. Carr became a commander on Skylab, America's first space station, in 1973–74.

Apollo 8's lead flight director, Cliff Charlesworth, died in

1991. The mission's other two flight directors, Glynn Lunney and Milton Windler, are retired.

At age ninety-five (as of 2019), Chris Kraft remains as sharp and feisty as ever. After a long and distinguished career at NASA, he retired as the director of the Johnson Space Center in 1982, then served as a consultant for several major corporations. He still has strong opinions on the space program, still remembers—vividly—watching Babe Ruth and Lou Gehrig hit home runs in the early 1930s.

George Low, the mastermind behind sending Apollo 8 to the Moon, became NASA's deputy administrator in late 1969. After retiring from the agency in 1976, he became the president of Rensselaer Polytechnic Institute. In 1984, he was awarded the Presidential Medal of Freedom for his work with America's space program. He died the next day, of cancer, at age fifty-eight.

In one of the last interviews of his life, in 2011, Neil Armstrong called Apollo 8 "an enormously bold decision" that catapulted the American space program forward. Harrison Schmitt, one of the two last people to set foot on the Moon as part of the crew of Apollo 17, said of the flight, "It was probably the most remarkable effort that the NASA team down here ever put together." When asked to compare Apollo 8 to his historic flight, astronaut Mike Collins said, "I think Apollo 8 was about leaving and Apollo 11 was about arriving, leaving Earth and arriving at the Moon. As you look back one hundred years from now, which is more important? I'm not sure, but I think probably you would say Apollo 8 was of more significance than Apollo 11." Astronaut Ken Mattingly said, "Of all of the events to participate in, you know, I was lucky because I could do Apollo 11 as well as 8 and then 13. But being part of Apollo 8, it made everything else anticlimactic."

For Chris Kraft, it was simple: "It took more courage to make the decision to do Apollo 8 than anything we ever did in the space program."

Weeks after the return of Apollo 8, Frank Borman went to work pumping gas at a local Gulf service station in Webster, Texas. He did it for free, along with his two sons, in exchange for use of the station's garage bay and lift, where the Bormans could work on their cars. The station was owned by a family friend, Toke Kobayashi, a man who also raised world-class tomatoes for sale to restaurants in Chicago.

Dressed in grimy jeans and a greasy T-shirt, Borman was almost unrecognizable to customers. One man who pulled in for gas began giving Borman a hard time, and for no good reason. Nearby, Borman's younger son, fifteen-year-old Ed, knew this was trouble—his father never started a fight but wouldn't take guff from anyone. Soon the men were near blows. Ed ran over and had to separate them, and it took every bit of his two-hundred-pound frame to do it. Even as Ed moved the belligerents apart, he found himself thinking: *This guy has no idea he's fighting with a man who just got back from the Moon.*

One day in 1969, as he was trying to figure out his post-NASA future, Borman's phone rang. On the line was a man who said, "At the behest of Ross Perot, I have invested a million dollars in your name." Perot wanted Borman to work for him by developing televised town hall forums in which the public could vote from home on issues of the day. The money, as the man said, had already been deposited. Borman was intrigued. Susan was not.

"You will not do that," she told her husband. "He'll own you. You don't know anything about this."

Borman gave back the money. He'd never made more than about thirty thousand dollars a year. Susan, who always received compliments on her fashionable wardrobe, would keep buying her clothes through secondhand stores and the Junior League thrift shop in Houston. More than ever, Frank was grateful for Susan's wisdom and good judgment. And he made a lifelong friend of Perot in the process.

After retiring from the Air Force in 1970, Borman joined Eastern Airlines, a position that required him to move to Miami. By now, his son Fred was at West Point, but seventeen-year-old Ed was a senior in high school and still living at home. Eastern wanted Borman to attend a three-month management program at Harvard Business School, a sure sign they had big plans for the former astronaut. Borman believed he needed Susan by his side in order to do his job well, so he asked his parents to stay with Ed for the year in Houston, and asked Susan to move with him to Miami. The prospect of leaving her son during his final year at home was deeply painful to Susan—her boys meant everything to her—but Frank needed her, so she packed up the car and went with him, much as she had done in the early years of their marriage. She was still plagued by depression, and her drinking had grown heavier. As always, she never showed any of that to Frank.

Borman rose quickly up the ranks at Eastern. While he impressed management, Susan struggled to adapt to a life in a new city, far away from her children and her friends. After his senior year, Ed followed his brother to West Point. For Susan, Florida felt emptier than ever.

In 1972, an Eastern Airlines jet crashed in an Everglades swamp on a flight from New York to Miami. The site was inaccessible by land, and rescue efforts were slow to mobilize. Unwilling to wait, Borman chartered a two-seat helicopter after midnight and flew with the pilot to search for the downed plane and survivors. The men found a tiny patch of solid ground on which to set down. Borman jumped from the chopper and into the waters of the swamp, which rose to his chest. All around, he heard moans and cries for help.

He worked to unpin victims from wreckage, helped the injured into arriving rescue helicopters, searched with a woman for her missing baby. Working a system of flashlights, he set up a local flight control, guiding choppers in and out of the scene. He departed on one of the last rescue craft out of the area, flying to the hospital to monitor the treatment of survivors. Of the 176 passengers and crew aboard Flight 401, 98 died in the accident.

Borman, who traveled constantly for work, was on assignment in New York in the fall of 1973 when he received a phone call telling him that Susan was very ill and advising him to return home immediately. It was past midnight, but he found an Eastern jet and jumped a ride on the empty plane. He had no idea what was wrong with his wife or how she was doing. It proved the longest and most helpless flight of his life. When he reached Susan's bedside the next morning, it became clear she'd had a nervous breakdown.

"I can't live like this, Frank," she told him. "I'm very sick but I'll do whatever it takes to get better."

Borman didn't know what to do. The doctor at Eastern Airlines did. "If you leave her here she's never going to get better, because she'll still be Mrs. Frank Borman of Eastern Airlines," he told Borman. The doctor had already made arrangements for Susan to go to the Institute of Living in Hartford, Connecticut, for treatment of her alcohol addiction, and for intensive psychotherapy.

On the flight to Hartford, Susan was nearly catatonic and didn't speak. At the treatment facility, the chief psychiatrist told Frank that Susan would need to be isolated and that Frank couldn't see or talk to her for a month. Frank couldn't imagine a more painful fate. Even at the Moon, she'd been with him every moment.

When doctors finally allowed him to visit, Frank found Susan much improved. They walked hand in hand across the grounds, and they talked. Frank felt a crushing guilt. All these years, he'd been selfish—mission had always come before family—and he'd never realized the toll this had taken on Susan. He, too, attended counseling sessions. He considered resigning from Eastern, changing his hard-charging ways. But he realized, with the doctor's help, that such a move would run counter to his DNA; it would do no one any good if he tried to become someone he was not. Instead, he promised to himself and to Susan: He would make more time for her, he would do more to communicate with her. And he swore to himself never to let anything like this happen again to the person he loved most.

Susan stayed for four months at the Institute of Living before returning home to Miami. From that day forward, neither she nor Frank touched alcohol again. Susan even brought home a friend from the facility, a young woman with addiction issues who'd been rejected by her family. Susan helped the woman find an apartment and a job, then counseled her for months until she'd settled in to the community. After that, Susan threw herself into volunteer work, helping organizations that fought drug abuse, an effort that would extend to a national scope in later years. Frank had never known a feeling of pride such as he felt for Susan in the months after she came home.

In May 1975, Borman was elected president and chief operations officer of Eastern Airlines. He was beloved by many in the company, from board members to pilots to mechanics. Often, he worked unloading baggage at the airport or checking engine parts on the tarmac, and he drove an old Chevy to work. In a later newspaper profile, another airline executive would say of him, "He kind of preceded all the 'excellence' books." Less than two years later, Borman became chairman of the board at Eastern, and he appeared in several of the company's television commercials. Even on TV, he couldn't help but talk straight. "Selling you a seat on Eastern Airlines isn't easy. It's not easy to sell you on any airline. You know, they're all pretty much the same," he said in one spot.

For several years under Borman, Eastern enjoyed record-setting profits. But labor difficulties, and the deregulation of the airline industry, caused a downturn in the company's business. Borman fought to right the ship, even making concessions that went against his instincts. For a time, the moves worked. But after a downturn in the economy, and new labor conflicts, Eastern was sold to new owners. After more than a decade at the helm, Borman resigned as the company's chairman in 1986.

No longer bound to Miami, the Bormans moved to Las Cruces, New Mexico, where their son Fred owned a car dealership. While there, Frank and Susan enjoyed one of the easiest and happiest stretches of their marriage. Frank served on corporate boards, invested in the car dealership with Fred, and stayed close to their other son, Ed, who'd become a helicopter test pilot. Frank did a lot of flying of small aircraft, still a foundational pleasure. Susan designed and rebuilt a home in the desert.

After more than a decade in New Mexico, Frank and Susan moved to Montana, following their son Fred, who'd purchased a cattle ranch. Frank continued to fly in Montana's big skies and attended air shows across the country with Susan. To this day, he thinks about a time in 1951 when, as an Air Force pilot, he ruptured an eardrum and was grounded permanently by order of the flight surgeon. Lying heartbroken in bed in the Philippines, he told Susan he'd leave the military and get a job as an aeronautical engineer. Susan had every reason to rejoice: She was a 21-year-old new mother with another baby on the way; Frank could earn a decent wage as a civilian; and the family could finally have a normal life, not one in which fighter pilots often died. Instead, Susan told her husband, "You will not do that. Flying means too much to you. You'll go see Major McGee and show him you can fly." Frank hardly knew what to say. But at Susan's urging, he asked Charles McGee, one of the original Tuskegee Airmen, to give him a shot. McGee didn't hesitate, putting Frank in an airplane and checking him out. When the flight surgeon found out that the legendary McGee had given his blessing, Frank was back in the cockpit. It's a story Borman seldom tells to others, but decades later he still can't get over what Susan did to save his career, and him.

People still recognize Frank in Montana sometimes. Some ask if he still looks up into the sky at the Moon and thinks about having gone there. He smiles, but tells the truth: "I suppose I do, but not often."

Shortly after Apollo 13's safe return in 1970, political heavy-weights in Wisconsin approached Jim Lovell about running as a Republican for United States Senate in his home state. He demurred, but that didn't stop Vice President Spiro Agnew, then President Nixon himself, from stepping in to press the recruiting effort. The election was just six months away; even with a war chest from the party, Lovell still had no structure, no party history. He passed. The decision pleased Marilyn. She knew Jim would make a fine senator but also knew that politics could tear a person's life and family apart. To her, going to the Moon had been a safer bet than going to Washington.

Lovell stayed at NASA for the next three years, working on early plans for the Space Shuttle and studying science applications for future space missions. The agency even sent him to a management program at Harvard, where he got his first taste of business.

All the while, the Navy had been calling, urging him to come back to the fleet, but promotion in that service didn't feel right to Lovell. He'd been a captain; the next step up was admiral. After more than a decade at NASA, he'd be competing against men who'd studied war and seen serious combat in Vietnam. To Lovell, they would always deserve promotion more than he. Given his management experience at Harvard, it seemed time to make the move into the private sector. When he retired from NASA and the Navy in 1973, no man had spent more time in space than Lovell, a total of 715 hours and five minutes, or just five hours short of a month.

As a private citizen, Lovell went to work for a tugboat company in Houston, where he became president and chief executive officer. The business proved lucrative, but after four years, Lovell found an even better opportunity, as president of a telecommunications company. It grew exponentially with deregulation in the industry, then was sold to Centel, a larger telecom corporation, where Lovell remained as an executive and board member until his retirement in 1991.

Four years later, Hollywood released a film version of *Apollo 13*, a book Lovell had coauthored about the rescue of that ill-fated mission. Tom Hanks played the lead role and the film became a hit. Suddenly, everyone wanted a piece of Lovell, and he did his best to oblige, touring the country and giving talks with his usual warmth and good humor. The new notoriety also helped him launch Lovell's of Lake Forest, a restaurant on Chicago's tony North Shore. Lovell packed the place with mementos from his NASA career, along with a giant mural behind the bar, titled *Steeds of Apollo*, that showed four horses galloping into the heavens. But the restaurant's real secret weapon was Lovell's son, Jay, its executive chef, who made Lovell's of Lake Forest a success.

On most evenings, the restaurant was packed with patrons. On the night of September 11, 2001, it was virtually deserted but for Jim Lovell himself. Dressed in his usual suit and tie, he stood alone in the corner of the basement bar, staring at the television, watching endless replays of terrorists bring down the towers at the World Trade Center in New York, an American hero not quite believing what had just happened to America.

In 2003, Jim and Marilyn Lovell celebrated the thirty-fifth anniversary of Apollo 8 with Frank and Susan Borman and Bill and Valerie Anders. The reunion was warm and friendly, but it seemed to Lovell that Frank was doing everything, even making lunch, while Susan seemed not to contribute much at all. But the whole event was so nice that Lovell didn't think much more about it.

Five years later, the couples met for the fortieth anniversary of Apollo 8. This time, it was even more evident to Lovell that something was amiss with Susan; she acted strangely and hardly talked, and Frank constantly helped her, even with small things.

Like his crewmates from Apollo 8, Lovell had flown private planes since retiring from NASA. Lately, he'd been piloting a twin-engine Cessna 421, able to make it from his home near Chicago to his other home in Texas—along with Marilyn, their dog, and luggage—in a slick five hours. But Marilyn had been growing increasingly concerned about her husband's flying alone. In the gentlest terms, she urged him to find a copilot or sell the airplane. In 2013, at age eighty-five, he sold. Not a day passes when he doesn't miss it.

In 2015, after a sixteen-year run, Lovell's of Lake Forest closed its doors, a victim of the economic downturn that began in 2008. Jay Lovell and his wife, Darice, opened a smaller, more casual place in nearby Highwood, Illinois. Jim and Marilyn eat there often.

At their home near Chicago, Jim and Marilyn lead a quieter life than they did during the NASA years, when they could hardly leave the house without attention. Now they enjoy their children and grandchildren, and a golden retriever they found at a rescue shelter—one that watches the *Apollo 13* movie whenever it plays on TV. For many years, Jim and Marilyn's daughter, Susan, wore the mink jacket that Jim had sent to Marilyn on Christmas Day 1968, while he was orbiting the Moon. But late in 2015, Marilyn asked for it back, if only for a Chicago winter shaping up to be colder than most.

Even now, Marilyn and Jim go for nightly walks near Lake Michigan. The shoreline makes for a perfect place to gaze up at the Moon, especially when it's full and just rising over the trees. In summer 2017, the International Astronomical Union formally recognized the name "Mount Marilyn" for the lunar mountain Lovell picked out for his wife during Apollo 8. Every once in a while during their walks, Jim shows Marilyn where it is on the Moon. Every once in a while, he thinks, *I've been there.*

By the time the last Apollo mission flew in late 1972, Bill Anders had been at the National Aeronautics and Space Council for almost four years. In that time, he'd worked on projects like Skylab (America's first space station), Viking (to put an unmanned spacecraft on Mars), and the Space Shuttle. He'd also lobbied successfully to include astronaut-geologist Harrison Schmitt on Apollo 17. To Anders, it made little sense for the agency to send only test pilots to land on the Moon; NASA needed someone who could expertly interact with and appreciate its geological wonders.

Before going to Washington, Anders had made a deal with NASA that permitted him to keep using the agency's T-38 airplanes for as long as he worked in government. The move allowed him to stay connected with high-performance jets and kept him sharp, just in case NASA extended Apollo into the future and changed its mind about keeping Anders in the command module pilot's seat.

Anders was still at the Space Council when he and Valerie welcomed their sixth child, daughter Diana. Soon after, he was appointed by President Nixon to the five-person Atomic Energy Commission (a move that made sense, given that Anders was a nuclear engineer), and two years later, in 1975, he was named the first chairman of the Nuclear Regulatory Commission. In six years of service in Washington, Anders had proved himself serious and nonpartisan. In 1976, a White House staffer reached out to Anders to see if he'd be interested in becoming an ambassador. Anders talked it over with Valerie, who expressed an interest in a country she'd found especially beautiful during a family visit. President Gerald Ford approved, and in 1976, Bill Anders became the United States ambassador to Norway.

While working in Oslo, Anders received a package from the International Astronomical Union, the organization in charge of naming surface features of planets. The IAU was pleased to announce it had named six lunar craters for Americans—the crews of Apollo 8 and Apollo 11—and six for Soviet cosmonauts. Included was a photograph of the craters. Anders was displeased: For starters, his crater wasn't the one that he'd named for his family during the Apollo 8 flight. Second, the astronauts' craters were in an area past the horizon that couldn't be seen. Third, the cosmonauts hadn't even been to the Moon (Anders joked that as atheists, they wouldn't even pass the Moon on their way to

Heaven). He called the organization and argued the astronauts' case, to no avail. The explorer's prerogative—to name the places one discovered—didn't seem to apply. Anders, however, wouldn't forget it. Years later, he would still be pushing the IAU to make things right. In 2018, fifty years after the flight of Apollo 8, the IAU officially named two lunar craters to honor the mission, "Anders Earthrise" and "8 Homeward." Both can be seen in the famous *Earthrise* photo taken by Anders during the flight.

Anders finally left government service in 1977, joining General Electric as vice president of its Nuclear Products Division. Two years later, the company sent him to the Advanced Management Program at Harvard Business School; when he returned, he became the general manager of GE's Aircraft Equipment Division, a multi-billion-dollar business. Over the next several years, Anders improved the division, all while absorbing the management style of GE's young new CEO, Jack Welch, who pushed to consolidate, simplify, move fast, and stay only in businesses in which his company could rank first or second in the industry.

After seven years at GE, Anders left to become chief operating officer of the aerospace and defense firm Textron. In 1989, he left that company to become vice chairman of General Dynamics, a major supplier of aircraft, tanks, and other weapons to the United States Department of Defense. By agreement, he would become the company's CEO a year later.

On paper, the move might have seemed crazy. After the fall of the Berlin Wall in 1989 signaled the coming end of the Cold War, defense contractors began to suffer, their wares no longer in name-your-price demand. General Dynamics seemed even worse off than its competitors; having amassed huge debt, it looked headed for bankruptcy. But Anders saw possibility in darkened clouds.

On becoming CEO, he instituted many of the principles he'd seen Welch use at GE. Among other moves, he sold off any part of the business in which General Dynamics couldn't be a market leader. He worked hard to change the corporate culture and become efficient, replacing most executives and many personnel, getting rid of waste endemic to the industry, and focusing on shareholder return. He even pitched in as a test pilot, flying the firm's F-16 fighter jets—until he sold off that part of the business, too.

The company's fortunes turned around fast. Billions of dollars flowed in, enough so that Warren Buffett purchased 16 percent of the company's stock—then gave Anders proxy to vote his shares. By the end of Anders's three-year term as CEO, he was a darling of Wall Street and, by many accounts, had saved General Dynamics. "After orbiting the Moon," one industry analyst said, "mundane business problems did not faze him."

Anders stayed on at General Dynamics for another year as chairman of the board, then retired from the company a wealthy man in May 1994. Soon after, he and Valerie fell in love with the natural beauty of Washington State, where they bought a house on the water and established the Anders Foundation, a philanthropic organization devoted to supporting education and the environment.

All the while, Anders kept flying. He'd already purchased a De Havilland Beaver airplane restored by his friend and former commander, Frank Borman, but what he truly envied was Borman's P-51 Mustang single-seat fighter-bomber, a workhorse from World War II and the Korean War. "If you ever find another, let me know," he told Borman.

Not long after, Anders found himself flying over Borman's home in New Mexico. He flipped open his cellphone and called to say hello.

"Hey, Anders," Borman said, "I found a Mustang for you—get your ass down here!"

The two men drove to inspect the plane.

"I'll buy it if you test it," Anders told Borman.

Borman, the old test pilot, put the plane through its paces. Anders wrote a check—and then had an idea.

He would start a museum dedicated to preserving—and

flying—historic military aircraft. The Mustang would be one of the first pieces in the family's Heritage Flight Museum, opened in 1996 in Burlington, Washington. It would also be the plane Anders flew in the 1997 Reno Air Races. At age sixty-four, he finished third in the silver race.

Along with Valerie and their children, Anders has helped run the foundation and the museum ever since. He still feels young; even in his mideighties, he's surprised to look in the mirror and find an elderly man looking back. He takes daily walks with Valerie. And he still flies, but not the warbirds anymore. Mostly, he takes a light two-seater, much like a Super Cub, over Washington skies. It's not a Mustang, or the F-89 he used to challenge Soviet bombers during the Cold War, but it's a hell of a lot better than not flying at all.

And he still cares about the environment. He knows that most people understand that Copernicus and Galileo were right, that the heavens do not revolve around Earth, but he wonders whether, down deep, any of us really believes it. By his estimation, human beings must think, in their reptilian brains, that Earth is flat and infinite; otherwise, they wouldn't treat it as badly as they do. To that end, the Anders Foundation continues to fight to protect the environment on the only planet any of us has.

Even after the fiftieth anniversary of Apollo 8, Anders thinks often of the view he got of his home planet from a distance of a quarter million miles. To him, Earth seemed staggeringly small, little more than a pinpoint in an infinite universe. That feeling has never left him. When he was young, Anders sometimes wondered about his place in the universe, and whether he was special. After the Moon, he didn't wonder about that anymore. After the Moon, he knew he wasn't special, and it brought him a kind of peace.

In April 2018, the crew of Apollo 8 joined the author in Chicago at the Museum of Science and Industry to celebrate the fiftieth anniversary of their mission. (Photos and video of the event can be seen at robertkurson.com/rocketmen.)

As of this printing all three of Apollo 8's astronauts were still married to their wives. They are the only crew that flew in either the Gemini or Apollo programs whose marriages all survived.

A few years ago, doctors diagnosed Susan Borman with Alzheimer's disease, for which there is presently no cure. (It was symptoms of this illness that Lovell had noticed in recent get-togethers with the Bormans.) Gradually, she lost her cognitive abilities; by 2015, she sometimes didn't recognize Frank or her sons, and needed a full-time care facility. From the moment she showed symptoms, Frank refused to leave her side, and has remained committed to her care ever since. Even at age ninety, he awakens at 5:30 A.M. to exercise, not for his own benefit, but to make sure he stays alive long enough to take care of Susan until the end of her days.

Every day, he visits her at the facility, where her room is decorated with photos of her family, and of the days when she and Frank had their adventures. In one, she is on the cover of a national magazine, looking more radiant than a Hollywood actress. "She was really beautiful," a nurse tells Frank. "Still is," Frank says.

It doesn't matter to Frank that Susan sometimes doesn't respond, or that she might not even know who he is. He still climbs into bed with her every day and lies next to her, still takes her to get her nails done at the beauty shop, still talks to her and tells her he loves her. He turns down invitations to travel or receive awards. He always eats nearby. Susan needs him.

He struggles to say it without tears but he must say it. "Susan is the best wife and best mother a person could ever hope to have. I was selfish. I was lucky."

And now he must return to her; it's visiting hours at the facility. As he walks from his truck toward the front door, it is clear he is on a mission, a new mission, the only mission more important than the Moon.

Acknowledgments

This book would not have been possible without the cooperation of the crew of Apollo 8. Frank Borman, Jim Lovell, and Bill Anders welcomed me into their homes for a series of interviews, then gave generously of their time for more than two years as I followed up with endless emails, calls, and questions. No matter how redundant or obvious or personal my queries, each man answered with kindness, patience, clarity, and humor. It is rare when pinnacle heroes measure up in private to their idealized public images, but the crew of Apollo 8 are the genuine article. I have never met three finer gentlemen.

I'm grateful to Marilyn Lovell and Valerie Anders for granting me interviews in their homes, and for delving deep into their personal lives and experiences. One cannot understand the story of Apollo 8, or realize the true strength behind the astronauts and their historic mission, without knowing these amazing women. I spoke briefly to Susan Borman while visiting in Montana, but she was too ill by that time to conduct interviews. Still, I feel like I came to know her through the memories and writings her husband and sons shared with me, and by being in her presence, surrounded by people who love her. I'm grateful to her family for allowing me that.

Many thanks to Chris Kraft, one of NASA's most legendary figures, for the two full days of interviews he gave me at his home in Houston. Kraft is a wonderful explainer, but by his eyes alone it was clear he still believed the decision to fly Apollo 8 to be the most courageous the space agency ever made. Others from NASA, including astronauts, engineers, and managers, granted me interviews in person and by phone, every one of them helpful to me in understanding both the Apollo 8 mission and the social, political, and scientific context in which it took place. For this, I'm thankful to Jerry Bostick, Mike Collins, Walt Cunningham, Gerry Griffin, Fred Haise, Glynn Lunney, Ken Mattingly, Milt Windler, and Al Worden.

I owe a debt of gratitude to Dr. David M. Harland, Dwayne Day, Frank O'Brien, J. L. Pickering, David Shomper, and Asif Siddiqi, who explained to me the complex workings of space flight and lunar missions, the history of the Space Race, and NASA's daring decision to fly Apollo 8. Warm thanks also to Clare Fentress and Andrew Billingsley for their superb research on this project; Charles Murray and Catherine Bly Cox for providing me with interviews they conducted decades earlier for their classic book *Apollo: The Race to the Moon;* Robert Feder, for his singular expertise on Walter Cronkite; Connie Moore at NASA; David Mosena and the staff at Chicago's Museum of Science and Industry; and to Ed Borman, Fred Borman, Dydia Delyser, Mark Foster, Jay Lovell, Susan Lovell, Diane Murphy, Sam Skinner, Pam Smith, and Mary Weeks for invaluable interviews and other contributions that helped me tell the story of Apollo 8.

A special thanks goes to Apollo historian W. David Woods, a world-class expert on NASA's lunar missions, and author of the book *How Apollo Flew to the Moon*. I discovered David while listening to a podcast about the Apollo program; I'd never heard someone explain technical matters so clearly and visually. I reached out to David at his home in Scotland and was thrilled when he agreed to consult with me. For the next two years, he answered questions, explained myriad concepts, and offered suggestions, all with a literary sensibility and the warmth and patience of an old friend. Writers sometimes need lucky breaks; one of my biggest came when I found David.

My publisher, Random House, continues to be like family to

me. I have been extraordinarily privileged to work with Kate Medina, my editor, since 2005, and have learned much from her about writing, storytelling, kindness, integrity, and decency; she remains one of my favorite people in the world. Anna Pitoniak, my other editor for this book, was a revelation. From the start, Anna urged me to view things at new angles and dig into lesser-known elements of the story, all with a gentle grace and deep insight into human nature. She pushed me even when I was convinced I'd gotten things right, and in every instance it made my work better. On top of it all, Anna is a wonderful writer and a lovely person. I was very fortunate to work with her.

At Random House, Tom Perry believed in this book from the start and has always believed in me; Sally Marvin has been my champion and friend since 2000; Gina Centrello has warmly supported me since I arrived at Random House; Dennis Ambrose has deftly guided all of my books through production with patience and good humor. Thanks, too, at Random House, to Aaron Blank, Maria Braeckel, Emily DeHuff, Melanie DeNardo, Andrea DeWerd, Joelle Dieu, Benjamin Dreyer, Toby Ernst, Erica Gonzalez, Anna Belle Hindenlang, Emily Kimball, Leigh Marchant, Mary Moates, and Bridget Piekarz. Carlos Beltrán and Edwin Tse designed the gorgeous cover for this book; Elizabeth Eno created its beautiful interior design.

My literary agent, Flip Brophy of Sterling Lord Literistic, and my film and television agent, Jon Liebman of Brillstein Entertainment Partners, are two of the best in the game, and have been part of my family for years; when they talk about business they are also talking about life, and I am better for all of our conversations. In Flip's office, Nell Pierce has been a joy to work with. In Jon's office, Nicki Beltranena has been incredibly insightful and hardworking, and helped to develop this book for the screen. Many thanks, also, to Brad Weston and Scott Nemes of Makeready for connecting early and intuitively with this book, for their passion, and for recognizing the story's potential for television.

I'm grateful to these people who read early drafts of *Rocket*

Men or have otherwise encouraged and supported my writing: Bill Adee, Dick Babcock, Andrew Beresin, Gabrielle Brussel, Andy Cichon, Josh Davis, Kevin Davis, Katelynd Duncan, Jonathan Eig, Joe Epstein, Robert Feder, Brad and Jane Ginsberg, David Granger, Peter Griffin, Rich Hanus, Elliott Harris, Miles Harvey, Neil Hirshman, John Jacobs, Jon Karp, Len and Pam Kasper, Jennie Lee, Melody Margolis, Gil Netter, Jason Steigman, Gary Taubes, Randi and Rob Valerious, Mark Warren, and Bill Zehme.

Thanks, also, to Ken Andre, Dr. Sanford Barr, Stu Berman, Mitch Cassman, Dr. Michael Davidson, Dr. Samuel Goldman, Jordan Heller, Dr. Nolen Levine, Mitch Lopata, Donna Moy, Scott Novoselsky, John Packel, Tracy Patis, Victor, Sally, and Virginia Reyes, Scott Rosenzweig, Kevin Sanders, Dr. Dan Schwartz, and Dan Warsh.

Ryan Holiday and Brent Underwood of Brass Check Marketing have been wonderful promoters for my books and have a very exciting company on their hands. Dr. Steven Tureff has been a blessing to my family for years. Joe Tighe was deeply kind and supportive during the writing of this book (and a keen reader, as well); I don't know how to thank him enough. It was too late to remember Rachel Harris Doxey in my last book, so I'm doing it here and sending love to her family and friends; we miss you.

My family always reads my work, cheers me on, and gives me wonderful (and honest) notes. Much love to Jane Glover (who read this book before anyone), Larry, Mike, and Sam Glover; the Wisniewski family; and to Ken, Steve, Carrie, and Chaya Kurson. Jane and Ken, my brother and sister, are better writers than I, but they are modest so I can live with it. My mom and dad, Annette and Jack Kurson, were the two best storytellers I've ever known; I hope I do them proud when I tell my kids stories on long drives and before bed, and in these books.

A special thanks goes to my friend Dave Shapson. I met him on the day I arrived as a freshman at the University of Wisconsin-

Madison in 1981. By 2:30 A.M., I was homesick and unable to sleep, and wandered into Open Pantry, where I found Dave (the only customer in the store) thumbing through magazines about science and space travel. We spoke about our mutual love of astronomy and our admiration for NASA, and I understood it when he lamented that so many great innovations in technology and space would come after our lifetimes. Dave has been among my closest friends ever since, one of the most unique and thoughtful people I've met. He is, at once, a master storyteller, a brilliant cook, a wonderful musician, a great listener, and a first-rate thinker. He sees more beauty in the world, and especially in the ordinary, than anyone I've known. I knew Dave would be excited when I undertook this book project, but couldn't imagine he would end up spending hundreds of hours to help me research, study, refine, and think it through. I never could have done this without him.

Finally, thanks to Amy, Nate, and Will Kurson. My sons always read my writing and talk through the architecture of my thinking. They find planets with me on smartphone apps when we look into the evening sky. They know things I don't know about the Moon. Amy is my best friend and soul mate. She gave more hours, and more love, to this book than most people give to their own careers, all while running her own full-time business and making a beautiful home. During the past three years, as I worked on this project, several people watched Amy and told me I was the luckiest man in the world, and I agree. My family is my everything. They guide me in the dark. They are my stars.

DIAGRAM OF APOLLO 8

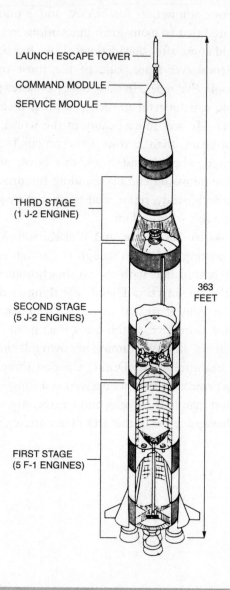

LAUNCH ESCAPE TOWER ——

COMMAND MODULE ——

SERVICE MODULE ——

THIRD STAGE
(1 J-2 ENGINE)

SECOND STAGE
(5 J-2 ENGINES)

363
FEET

FIRST STAGE
(5 F-1 ENGINES)

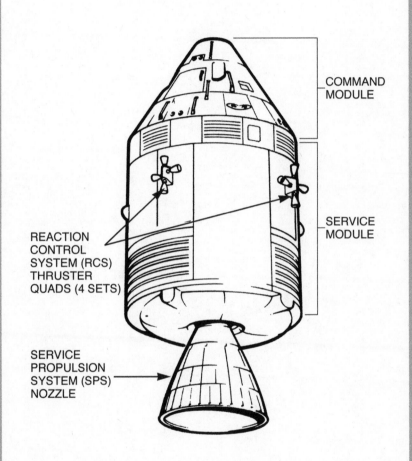

COMMAND
MODULE

SERVICE
MODULE

REACTION
CONTROL
SYSTEM (RCS)
THRUSTER
QUADS (4 SETS)

SERVICE
PROPULSION
SYSTEM (SPS)
NOZZLE

Author's Note

In late 2014, I took some friends to the Museum of Science and Industry in Chicago, my hometown. It had been a few years since I'd seen *U-505*, the German U-boat that is one of the centerpiece attractions in the museum, and a perfect match for the submarine I wrote about in my first book, *Shadow Divers*.

On the way out, we came across a space capsule, about ten feet tall by thirteen feet wide. It appeared to be scarred from its journey, wherever it had gone, and its open hatch revealed three cramped seats and a universe of controls inside. Kids circled around the spacecraft, which looked at once to have come from the past and the future.

A nearby placard announced that this was the command module of Apollo 8, which had carried the first men ever—Frank Borman, James Lovell, and William Anders—to the Moon. I knew almost nothing about that mission. Like many, I was much more familiar with the story of Apollo 11, man's first lunar landing, and of Apollo 13, when an explosion on board the spacecraft nearly resulted in tragedy. A few weeks later, I got around to reading about Apollo 8.

What I found was one of the most incredible stories in American history.

It had everything—daring, adventure, risk-taking, a race against time that came down to the final hours, an existential battle against a magnificent adversary. It blended cutting-edge science and technology with the eternal human yearning to explore. It told of the power of three unbreakable women and the love of children and family, of America's ability to do the impossible when pushed to its limits, of the moment when mankind first reached the place that had called to it for eternity—the Moon. It told of how three men lived extraordinary lives after becoming the first ever to leave the world. It was even a Christmas story.

The more I read about the odyssey of Apollo 8, the more startling it seemed that so little had been written about it. *This is the best space story of them all*, I thought, and I wasn't the only one. Early in my research, I came across interviews with the late Neil Armstrong, the first man to set foot on the Moon (and a backup crew member for Apollo 8). He remembered how excited all the astronauts and NASA personnel had been for Apollo 8, how it changed the course of the entire American space program. "It was an *enormously* bold decision," he told an interviewer on film. It was the way he said the word "enormously" that stayed with me. In ways, it sounded like Armstrong thought Apollo 8 to be an even bigger leap for mankind than landing on the Moon.

Other astronauts and NASA personnel said as much directly. Several called Apollo 8 the riskiest and most thrilling of all the Apollo missions. Few remembered having dry eyes as Borman, Lovell, and Anders spoke to the world on Christmas Eve as they circled the Moon. All of them—along with billions of others around the world, more than any than had ever listened to a human voice at once—remembered what these three astronauts said.

As I pushed deeper into researching Apollo 8, I found another story, one with striking parallels to life in America today. Apollo 8 flew at the end of 1968, one of the most terrible and divisive years in the country's history. Assassinations, riots, war, and other events split the country and turned neighbor against neighbor, Republican against Democrat, young against old. When Apollo 8 flew at the end of December, it looked like nothing could heal a nation so badly wounded from the inside.

Nearly fifty years later, the United States seemed torn apart again. As candidates launched their presidential campaigns in 2015, the country stood divided by a world of political and cultural differences, some of which manifested in violence or ugly public displays. Many people had never seen their country so fractured, their fellow citizens so furious with one another, and it only got worse as the election approached and then a new president was elected. But to those old enough to remember, it all looked so much like 1968.

There was one significant difference between 1968 and modern-day America, however. In 1968, there was Apollo 8. When Borman, Lovell, and Anders returned from the Moon, few could argue—no matter their age or political leaning or background—that they hadn't seen something important and beautiful happen, that these three men had helped the country, and the world, to heal. So far, there has been no Apollo 8 for our time.

I knew right away that I wanted to tell the story of Apollo 8. But I also knew that I couldn't do the story justice without interviewing the three crew members. At the time, Borman and Lovell were 87 years old, Anders was 83. I wasn't certain that any of them would want to talk to me.

I found Lovell's email address and wrote to him. A few days later, his assistant called and said he would be pleased to meet with me at his office, just a 15-minute drive from my home. When I arrived, Lovell told me that years earlier he'd listened to *Shadow Divers* as an audiobook and found himself so engrossed he'd circled in the parking lot at his office, unwilling to leave the car until a chapter had ended. I don't know if I've ever had a thrill like that—envisioning an Apollo astronaut going into orbit so he could finish my book.

For hours, Lovell told me about Apollo 8, his childhood growing up in Milwaukee, attending the University of Wisconsin-Madison. Even better, he agreed to meet with me again, and passed along contact information for Borman and Anders. Soon

I was in Montana, spending several days with Borman, and in Washington State, on an extended stay to interview Anders and his wife, Valerie, then back to Montana. In Chicago, I began a series of interviews with Lovell and his wife, Marilyn, at their home, where I got to see astonishing space memorabilia and pet their golden retriever, Toby, while asking my questions.

All of the men, and their wives, were extraordinarily generous with their time. (Susan Borman was too ill during my visits to Montana to sit for interviews.) And all of them were excellent storytellers, forthcoming and vivid in their recall. It was lucky for me that each of the astronauts understood how to explain even the most technical information in a way that a layperson could understand.

As informative as the astronauts were during my visits, they were equally down to earth and kind. Borman and Anders each took me for a ride in their small airplanes; I don't know that I've ever experienced a feeling like I got while being flown over the American countryside by the first men to see Earth from another world. (By the time I met Lovell, he'd given up flying for health reasons.) And it was moving to hear, through my headset, the admiration paid to Borman by a young flight controller in Montana, who seemed in awe that he was giving clearance to the first man who'd ever reached the Moon.

On my behalf, Borman called Chris Kraft, one of the titans of NASA, and a man as responsible as anyone for the success of the American space program. A few weeks later, I was in Houston meeting with Kraft, already in his nineties and possessing enough energy to go back and run Mission Control (which he invented). Kraft gave me two full days of interviews, and helped me reach other NASA veterans who lived in and around Houston. My week there was immensely rewarding in the company of these pioneers.

On my last day in Houston, I took my thirteen-year-old son to the Johnson Space Center to see the Saturn V rocket, still the most powerful machine ever built. Everyone I'd spoken to, from astronauts to NASA personnel to technical experts, warned that the immensity of the Saturn V couldn't quite be described, that one had to be in its presence to believe it. We paid our admission, admired the Space Shuttle, took turns in a simulator that twisted and shook. Then, we found the rocket.

It was laid on its side, 363 feet long end-to-end, bursting out of its own building. A decade-and-a-half into the twenty-first century, it's near impossible to find a piece of technology from the past that can impress a thirteen-year-old who owns an iPhone, an Xbox, and a quad-core computer. My son stood beside the five F-1 engines at the base of the rocket. He didn't look at his phone or check his texts. He didn't take a picture. He just kept staring at the nozzles of these engines, each more than twelve feet tall, and after staying still for several minutes, he asked if we could stay some more.

I made one more trip to Montana after that, to see Borman. After several days, we wrapped up by going to dinner at one of his favorite barbeque restaurants. We'd been seated only a few minutes when a bartender rushed up to him holding a newspaper. "Frank, the paper says you're one of Montana's most famous residents!" she said. The woman couldn't have been more than twenty years old. It wasn't clear she had any idea who Borman was or why he was famous, but he smiled and said "Thank you very much" all the same.

After the meal, Borman dropped me off at my hotel, then went to visit his wife at the nursing home where she lives. As he drove away, it seemed to me strange—I felt I'd come to know Susan as well as I had Frank, despite having met her for just a few minutes, despite the fact that she had been too ill to speak. When I returned home and transcribed the tapes of my interviews, I understood why. Borman spoke of Susan constantly; there didn't seem an aspect of his life he could explain without discussing how much she meant to him or how much he loved her. I'd heard the same from Lovell and Anders about their wives. When I discovered that Apollo 8 was the only crew in which all the marriages survived (astronaut careers were notoriously hard on marriages) it didn't surprise me. In a singularly beautiful story, it seemed only fitting that the first men to leave Earth considered home to be the most important place in the universe.

A Note on Sources

The heart and soul of this book come from extensive interviews I conducted with Frank Borman, Jim Lovell, and Bill Anders, the three astronauts who flew on Apollo 8. I met with each man over the course of several days at his home, and followed up repeatedly by phone and email, compiling dozens of hours of recorded conversation. Despite their ages (Borman and Lovell were born in 1928, Anders in 1933), all three of them seemed to have as much energy, and were just as sharp, as when they became the first men ever to fly to the Moon fifty years ago. It was I who often struggled to keep up with them.

Equally important to the book are the interviews I conducted with two of the astronauts' wives, Marilyn Lovell and Valerie Anders. Both sat for several hours over many days, and made me feel welcome in their homes. By the time I undertook this project, Susan Borman was too ill to talk, but I met her, and Frank supplied me with much background information from a private journal he kept about her life and times. Quotes from Susan that appear in the book are from interviews she did when she was well, along with accounts from Frank, her two sons, and others who knew her.

It's difficult to imagine having written this book without the extraordinary generosity of Chris Kraft, one of the most important figures in NASA's history. At age ninety-one, Kraft welcomed me to his home in Houston for two full days of interviews about the flight of Apollo 8 and the bold series of decisions that led up to it. Like the astronauts, Kraft was in peak form and recalled details and events as clearly—and cared about them as much—as if they'd happened yesterday.

During the course of reporting for this book, I also interviewed several other astronauts and NASA personnel who were ringside for Apollo 8. I would have loved to talk to more of them, but by the time I began work on the project, many had already passed away. Still, I was lucky. Over the decades, NASA and other organizations had the foresight to record interviews with a great number of people involved in the American space program. I made use of eighty or more of these oral histories, including those with astronaut Neil Armstrong (who was on the backup crew for Apollo 8) and NASA giants including Robert Gilruth, George Mueller, Samuel Phillips, and James Webb. In addition, I benefited greatly from the generosity of Charles Murray and Catherine Bly Cox, who supplied me with more than a dozen transcripts of interviews they conducted in the 1980s for their classic book *Apollo: The Race to the Moon*. Among other gems, it was their interview with Judy Wyatt that provided the detail about George Low's subjects and verbs agreeing; and it was their interviews with Mission Control personnel that revealed that some NASA managers—only half jokingly—remained concerned that Apollo 8 might smash into the Moon even as the spacecraft closed in on the lunar surface.

Every aspect of the Apollo 8 story—the conception, planning, and execution of the mission, the American space program, the Space Race between the United States and the Soviet Union, the Cold War—was exhaustively documented as it unfolded. Even papers once secret have now been declassified. I benefited from all of this.

For my purposes, the single most important documentary source on the flight of Apollo 8 was the *Apollo 8 Flight Journal*, a Web-based transcript of the available recordings from the mission, along with corrections and a running series of astonishingly clear commentaries and explanations. The *Flight Journal* was created by David Woods, with help from Frank O'Brien, based on the *Apollo Lunar Surface Journal* by Eric Jones, who saw the power of the Internet to host these massive, ever-changing documents. Hosted with the kind assistance of the NASA History Division, the *Apollo 8 Flight Journal* was essential to my understanding—on a minute-by-minute basis—of the historic six-day mission. I was also extremely fortunate to work with David Woods during the research and writing of my book, as he helped to clarify and confirm my understanding of the flight. The *Apollo 8 Flight Journal* can be accessed online at history.nasa.gov/afj /ap08fj/index.html. In addition, I benefited greatly by my consultation with space historian Dr. David M. Harland, who has written extensively about the Apollo program, including a masterwork, *Exploring the Moon: The Apollo Expeditions.*

Also essential to my understanding of the flight—and to my sense of being there—were two online series of videos about Apollo 8 that included film, audio transmissions between the spacecraft and Mission Control (with time markers), live television broadcasts, animations, press briefings, and other aspects of the mission. Perhaps the most useful was this series of forty-two videos that covers the entire mission: youtube.com/playlist?list= PLC1yaZz2qeGogsUbODzdA0-iJ8Qtb6kEB. For another excellent series, see this collection of sixty-one videos of CBS News coverage of Apollo 8: youtube.com/playlist?list=PLwxFr1zAEfo lhvY0z_lSMAQFuZDqatinX.

As one might expect with an Apollo mission, NASA produced an immense cache of primary source material, including thousands of documents from the agency covering every aspect of Apollo 8, from conception to construction to training to the flight itself and its aftermath. Some of it was far too technical for the purposes of this book, but much of it proved invaluable. Among the treasures I mined from this wonderland of documents: flight plans (which look like hieroglyphics to the untrained eye, but once deciphered become a bible of the mission);

NASA memos and analyses; mission rules and procedures; flight evaluations; checklists; public affairs commentaries; transcripts of onboard voice transmissions; press briefings; crew debriefings; photographs and visual observations; and chronologies. I also repeatedly turned to a 448-page document compiled by NASA titled "Astronautics and Aeronautics, 1968—Chronology on Science, Technology, and Policy," a day-by-day account of the American space program in one of its most critical years.

Whenever there was a discrepancy in a version of events, I used my best efforts to present the most likely and clearest account. Quotations from the flight of Apollo 8 are taken from mission transcripts; in a few instances, when the astronauts were not broadcasting their conversations, I have presented their dialogue as they recounted it for me. Other dialogue in the book comes from the sources listed in this note. Occasionally, I assembled dialogue to reflect stories I was told or that I researched.

I sent a draft of this book to Frank Borman, Jim Lovell, and Bill Anders to review for factual accuracy. I did not seek, nor did the astronauts offer, editorial changes. I'm grateful to them for their time and attention to detail, and for helping to make the manuscript as accurate as possible.

In addition to primary sources generated by NASA, I consulted hundreds of books, magazines, newspaper articles, websites, documentaries, films, audio recordings, photographs, and podcasts. I found value in nearly all of them, but I returned to the following time and again for their excellence, clarity, and breadth of information:

BOOKS

Chaikin, Andrew. *A Man on the Moon: The Voyages of the Apollo Astronauts.* New York: Penguin, 2007.

French, Francis, and Colin Burgess. *In the Shadow of the Moon: A Challenging Journey to Tranquility, 1965–1969.* Lincoln and London: University of Nebraska Press, 2007.

Kraft, Christopher. *Flight: My Life in Mission Control*. New York: Dutton, 2001.

Murray, Charles, and Catherine Bly Cox. *Apollo: The Race to the Moon*. New York: Simon and Schuster, 1989.

Siddiqi, Asif. *Challenge to Apollo: The Soviet Union and the Space Race, 1945–1974*. alc Books, 2015.

Woods, W. David. *How Apollo Flew to the Moon*. 2nd ed. New York: Springer, 2011.

Zimmerman, Robert. *Genesis: The Story of Apollo 8*. New York: Basic Books, 1998.

WEBSITES

nasa.gov

thespacereview.com

airandspace.si.edu

airspacemag.com

space.com

collectspace.com

MAGAZINES

Aviation Week & Space Technology (all 1968 issues)

Life (which had exclusive access to the astronauts and their families)

DOCUMENTARIES AND VIDEOS

"Apollo 8 Reunion 2008—An Evening with the Apollo 8 Astronauts" (Annual John H. Glenn Lecture Series)

"Apollo 8 Reunion 2009"

Cold War (24-episode television documentary, originally broadcast in the United States on CNN)

Race to the Moon: The Daring Adventure of Apollo 8 (American Experience—PBS / Indigo Studios 2005)

svs.gsfc.nasa.gov//4129 (a brilliant animated explanation of Earthrise, narrated by Andrew Chaikin)

youtube.com/watch?v=Vn00BvWwke0 (launch of Apollo 8)

Please check my website (robertkurson.com) for a more comprehensive list of sources used to research and write this book, along with copies of NASA documents and links to video, audio, and other multimedia sources.

Index

PHOTO: © MATT FERGUSON

ROBERT KURSON earned a bachelor's degree
in philosophy from the University of Wisconsin and a law degree from Harvard Law
School. His award-winning stories have appeared in *Rolling Stone, The New York Times
Magazine,* and *Esquire,* where he was a con-
tributing editor. He is the author of four
New York Times bestsellers: *Shadow Divers,*
the 2005 American Booksellers Association's
nonfiction Book Sense Book of the Year;
Crashing Through, based on Kurson's 2006
National Magazine Award–winning profile
of the blind speed skier, CIA agent, inven-
tor, and entrepreneur Mike May in *Esquire;*
Pirate Hunters; and *Rocket Men,* which tells
the story of the Apollo 8 mission to the
Moon. He lives in Chicago.

RobertKurson.com
Twitter: @robertkurson